Collins

Student Support Materials for

Edexcel A Level Maths

Core 4

Authors: John Berry and Sue Langham

William Collins' dream of knowledge for all began with the publication of his first book in 1819. A self-educated mill worker, he not only enriched millions of lives, but also founded a flourishing publishing house. Today, staying true to this spirit, Collins books are packed with inspiration, innovation and practical expertise. They place you at the centre of a world of possibility and give you exactly what you need to explore it.

Collins. Freedom to teach.

Published by Collins
An imprint of HarperCollins*Publishers*
77–85 Fulham Palace Road
Hammersmith
London
W6 8JB

10 9 8 7 6 5 4 3 2 1

ISBN: 978-0-00-747604-6

British Library Cataloguing in Publication Data. A Catalogue record for this publication is available from the British Library.

Commissioned by Lindsey Charles and Emma Braithwaite
Project managed by Lindsey Charles
Edited and proofread by Susan Gardner
Reviewed by Stewart Townend
Design and typesetting by Jouve India Private Limited
Illustrations by Ann Paganuzzi
Index compiled by Michael Forder
Cover design by Angela English
Production by Simon Moore

Printed and bound in Spain by Graficas Estella

This material has been endorsed by Edexcel and offers high quality support for the delivery of Edexcel qualifications.

Edexcel endorsement does not mean that this material is essential to achieve any Edexcel qualification, nor does it mean that this is the only suitable material available to support any Edexcel qualification. No endorsed material will be used verbatim in setting any Edexcel examination and any resource lists produced by Edexcel shall include this and other appropriate texts. While this material has been through an Edexcel quality assurance process, all responsibility for the content remains with the publisher.

Copies of official specifications for all Edexcel qualifications may be found on the Edexcel website - www.edexcel.com

Browse the complete Collins catalogue at: www.collinseducation.com

Acknowledgements
The publishers wish to thank the following for permission to reproduce photographs. Every effort has been made to trace copyright holders and to obtain their permission for the use of copyright material. The publishers will gladly receive any information enabling them to rectify any error or omission at the first opportunity.

Cover image: Abstract Glass Side of Business Building
© Vladitto | Dreamstime.com

MIX
Paper from responsible sources
FSC® C007454

FSC™ is a non-profit international organisation established to promote the responsible management of the world's forests. Products carrying the FSC label are independently certified to assure consumers that they come from forests that are managed to meet the social, economic and ecological needs of present and future generations, and other controlled sources.

Find out more about HarperCollins and the environment at
www.harpercollins.co.uk/green

Welcome to Collins Student Support Materials for Edexcel A level Mathematics. This page introduces you to the key features of the book which will help you to succeed in your examinations and to enjoy your maths course.

The chapters are organised by the main sections within the specification for easy reference. Each one gives a succinct explanation of the key ideas you need to know.

Examples and answers

After ideas have been explained the worked examples in the green boxes demonstrate how to use them to solve mathematical problems.

Method notes

These appear alongside some of the examples to give more detailed help and advice about working out the answers.

Essential ideas

These are other ideas which you will find useful or need to recall from previous study.

Exam tips

These tell you what you will be expected to do, or not to do, in the examination.

Stop and think

The stop and think sections present problems and questions to help you reflect on what you have just been reading. They are not straightforward practice questions - you have to think carefully to answer them!

Practice examination section

At the end of the book you will find a section of practice examination questions which help you prepare for the ones in the examination itself. Answers with full workings out are provided to all the examination questions so that you can see exactly where you are getting things wrong or right!

Notation and formulae

The notation and formulae used in this examination module are listed at the end of the book just before the index for easy reference. The formulae list shows both what you need to know and what you will be given in the exam.

Contents

Partial fractions

In Core 3 you developed the skill of adding two or more algebraic fractions to give a single fraction in the following way.

Step 1: Given $\frac{1}{2x-1}+\frac{1}{x+2}$ find the lowest common denominator of the two fractions by multiplying the two denominators:

$(2x-1)(x+2)$

Step 2: Write the two fractions with this common denominator:

$$\frac{1}{2x-1}+\frac{1}{x+2}\equiv\frac{1(x+2)}{(2x-1)(x+2)}+\frac{1(2x-1)}{(x+2)(2x-1)}$$

Step 3: Collect together the two fractions as they now have the same denominator:

$$\equiv\frac{(x+2)+(2x-1)}{(2x-1)(x+2)}$$

$$\equiv\frac{3x+1}{(2x-1)(x+2)}$$

Essential notes

The lowest common denominator is the algebraic expression of the lowest degree (or power) which has $(2x-1)$ and $(x+2)$ as factors.

We now consider the method needed for the reverse process. We start with a single algebraic fraction and write it as the sum of two other fractions. These two fractions are called the **partial fractions** of the original algebraic fraction. Partial fractions are particularly useful when differentiating algebraic fractions as the following example illustrates.

Example

Differentiate $y=\frac{3x+1}{(2x-1)(x+2)}$

Answer

Step 1: Write the given fraction in terms of partial fractions as explained above:

$$y=\frac{3x+1}{(2x-1)(x+2)}\equiv\frac{1}{2x-1}+\frac{1}{x+2}$$

Step 2: Rewrite each partial fraction using the rule of indices:

$$y=\frac{1}{2x-1}+\frac{1}{x+2}\equiv(2x-1)^{-1}+(x+2)^{-1}$$

Step 3: Differentiate each partial fraction using the chain rule:

$$\frac{dy}{dx}=(-1(2x-1)^{-2}\times 2)+(-1(x+2)^{-2}\times 1)$$

Step 4: Rewrite using the rule of indices:

$$\frac{dy}{dx}=\frac{-2}{(2x-1)^2}+\frac{-1}{(x+2)^2}=-\frac{2}{(2x-1)^2}-\frac{1}{(x+2)^2}$$

Method notes

An alternative method of differentiating would be to use the quotient rule on $\frac{3x+1}{(2x-1)(x+2)}$ with $u=(3x+1)$ and $v=(2x-1)(x+2)$.

Method notes

The chain rule $\frac{dy}{dx}=\frac{dy}{du}\times\frac{du}{dx}$ was covered in Core 3.
It is used to differentiate a composite function of x.

This chapter explores various methods for writing single fractions in terms of partial fractions.

Fractions with linear factors in the denominator

Example

Express $\dfrac{2x + 1}{(x - 1)(x + 2)}$ in partial fractions.

Answer

Step 1: Let $\dfrac{2x + 1}{(x - 1)(x + 2)} \equiv \dfrac{A}{(x - 1)} + \dfrac{B}{(x + 2)}$

where A and B are constants which are to be evaluated.

Step 2: Write the two partial fractions with a common denominator of $(x - 1)(x + 2)$:

$$\frac{A}{(x - 1)} + \frac{B}{(x + 2)} \equiv \frac{A(x + 2)}{(x - 1)(x + 2)} + \frac{B(x - 1)}{(x + 2)(x - 1)}$$

Step 3: Collect together the two fractions as they have the same denominator:

$$\frac{2x + 1}{(x - 1)(x + 2)}$$

$$\equiv \frac{A}{(x - 1)} + \frac{B}{(x + 2)}$$

$$\equiv \frac{A(x + 2) + B(x - 1)}{(x - 1)(x + 2)}$$

Step 4: The denominators of the fractions in step 3 are equal so the numerators must also be equal:

$2x + 1 \equiv A(x + 2) + B(x - 1)$

This is an identity which means that it is true for all values of x.

If we now choose suitable values of x to substitute into this identity it will enable us to evaluate A and B.

Step 5: Let $x = 1$ in the identity $2x + 1 \equiv A(x + 2) + B(x - 1)$

$\Rightarrow 3 = 3A + 0B$ so $A = 1$

Step 6: Let $x = -2$ in the identity $2x + 1 \equiv A(x + 2) + B(x - 1)$

$\Rightarrow -3 = 0A + (-3)B$ so $B = 1$

Step 7: Rewrite the original fraction using the values of A and B hence:

$$\frac{2x + 1}{(x - 1)(x + 2)} \equiv \frac{1}{(x - 1)} + \frac{1}{(x + 2)}$$

Method notes

This fraction has **two** linear factors in the denominator. We can write this as the sum of **two** partial fractions by using the two denominator factors.

Essential notes

If two fractions are equal to each other and the denominators of those fractions are equal to each other then the numerators of those fractions must also be equal to each other.

Method notes

The most suitable x-values to choose to find A and B are found by making each of the brackets which multiply A and B in the identity zero. Therefore in

$2x + 1 \equiv A(x + 2) + B(x - 1)$

make $(x + 2) = 0 \Rightarrow x = -2$

and make $(x - 1) = 0 \Rightarrow x = 1$

Method notes

This fraction has **one quadratic function** in the denominator which we must factorise into **two** linear factors. These linear factors are then used for writing the original fraction as the sum of **two** partial fractions.

Example

Express $\dfrac{2}{(2x^2+5x-3)}$ in partial fractions.

Answer

Step 1: Factorise the denominator $2x^2+5x-3 \equiv (2x-1)(x+3)$.

Step 2: Rewrite the original fraction using the result in step 1:

$$\frac{2}{(2x^2+5x-3)} \equiv \frac{2}{(2x-1)(x+3)}$$

Step 3: Let $\dfrac{2}{(2x-1)(x+3)} \equiv \dfrac{A}{(2x-1)} + \dfrac{B}{(x+3)}$ where A and B are constants.

Step 4: Rewrite $\dfrac{A}{(2x-1)} + \dfrac{B}{(x+3)}$ with their common denominator $(2x-1)(x+3)$:

$$\frac{A}{(2x-1)} + \frac{B}{(x+3)} \equiv \frac{A(x+3)+B(2x-1)}{(2x-1)(x+3)}$$

Step 5: Equate the numerators from step 4:

$$2 \equiv A(x+3) + B(2x-1)$$

Step 6: Choose a suitable x-value to evaluate B:

$(x+3) = 0 \Rightarrow x = -3$

So in the identity from step 5 when $x = -3$:

$$2 = 0A + (-7)B$$

$$\Rightarrow B = -\frac{2}{7}$$

Step 7: Choose a suitable x-value to evaluate A:

$(2x-1) = 0$ so let $x = \dfrac{1}{2}$ in the identity from step 5 when $x = \dfrac{1}{2}$:

$$2 = \frac{7}{2}A + (0)B$$

$$\Rightarrow A = \frac{4}{7}$$

Step 8: Rewrite the original fraction using the values of A and B:

$$\frac{2}{(2x^2+5x-3)}$$

$$\equiv \frac{4}{7(2x-1)} - \frac{2}{7(x+3)}$$

Method notes

Be careful when substituting the numerical values of A and B back into step 3.

Check that you have used the correct letter value for each fraction.

Example

Express $\dfrac{2x+1}{(x-1)(x+1)(x+2)}$ in partial fractions.

Answer

Step 1: Let $\dfrac{2x+1}{(x-1)(x+1)(x+2)} \equiv \dfrac{A}{(x-1)} + \dfrac{B}{(x+1)} + \dfrac{C}{(x+2)}$

where A, B and C are constants.

Step 2: Rewrite $\dfrac{A}{(x-1)} + \dfrac{B}{(x+1)} + \dfrac{C}{(x+2)}$ with their lowest common denominator which is $(x-1)(x+1)(x+2)$:

$$\frac{A}{(x-1)} + \frac{B}{(x+1)} + \frac{C}{(x+2)}$$

$$\equiv \frac{A(x+1)(x+2)}{(x-1)(x+1)(x+2)} + \frac{B(x-1)(x+2)}{(x-1)(x+1)(x+2)}$$

$$+ \frac{C(x-1)(x+1)}{(x-1)(x+1)(x+2)}$$

$$\Rightarrow \frac{A}{(x-1)} + \frac{B}{(x+1)} + \frac{C}{(x+2)}$$

$$\equiv \frac{A(x+1)(x+2) + B(x-1)(x+2) + C(x-1)(x+1)}{(x-1)(x+1)(x+2)}$$

we know from step 1 that $\dfrac{2x+1}{(x-1)(x+1)(x+2)}$

$$\equiv \frac{A(x+1)(x+2) + B(x-1)(x+2) + C(x-1)(x+1)}{(x-1)(x+1)(x+2)}$$

Step 3: Equate the numerators from step 2:

$$2x+1 \equiv A(x+1)(x+2) + B(x-1)(x+2) + C(x-1)(x+1)$$

Step 4: Choose suitable x-values to evaluate A, B and C:

$(x-1) = 0 \Rightarrow x = 1$

$(x+1) = 0 \Rightarrow x = -1$

$(x+2) = 0 \Rightarrow x = -2$

Step 5: Substitute these x-values into the identity in step 3:

when $x = 1$, $3 = 6A + (0)B + (0)C \quad \Rightarrow A = \dfrac{1}{2}$

when $x = -1$: $-1 = (0)A + (-2)B + (0)C \quad \Rightarrow B = \dfrac{1}{2}$

when $x = -2$: $-3 = (0)A + (0)B + 3C \quad \Rightarrow C = -1$

Step 6: Rewrite the original fraction using the values of A, B and C

hence $\dfrac{2x+1}{(x-1)(x+1)(x+2)} \equiv \dfrac{1}{2(x-1)} + \dfrac{1}{2(x+1)} - \dfrac{1}{(x+2)}$

Method notes

This fraction has **three linear factors** in the denominator.

We can write this as the sum of **three partial fractions** by using the three denominator factors.

Essential notes

The degree or power of a polynomial was covered in Core 1

The degree is the highest power in the polynomial.

$x^3 + 2x$ is of degree three, $2x^2 + 3$ is a quadratic expression of degree two,

$7x + 5$ is a linear expression of degree one.

$3 = 3(x^0)$ is of degree 0

All the partial fractions in previous examples had a linear denominator and a constant numerator.

Such algebraic fractions are called **proper** algebraic fractions because the numerator was of lower degree than the denominator.

The fractions $\frac{(2x - 5)}{(x^2 + 3)}, \frac{(3x + 7)}{(x^2 + 1)}$ are also examples of proper algebraic fractions.

We now consider examples where the partial fractions have a quadratic denominator which does not factorise.

Fractions with quadratic factors in the denominator

Method notes

If there is a quadratic factor in the denominator of the fraction which will not factorise we write a linear expression in the numerator. This ensures that the partial fraction is proper. In this example we choose $(Bx + C)$ as the linear numerator where B and C are constants.

Example

Express $\frac{19}{(x - 4)(x^2 + 3)}$ in partial fractions.

Answer

Step 1: Let $\frac{19}{(x - 4)(x^2 + 3)} \equiv \frac{A}{(x - 4)} + \frac{(Bx + C)}{(x^2 + 3)}$ where A, B and C are constants.

Step 2: Rewrite $\frac{A}{(x - 4)} + \frac{(Bx + C)}{(x^2 + 3)}$ with their lowest common denominator $(x-4)(x^2+3)$:

$$\frac{A}{(x - 4)} + \frac{(Bx + C)}{(x^2 + 3)} \equiv \frac{A(x^2 + 3) + (Bx + C)(x - 4)}{(x - 4)(x^2 + 3)}$$

we know from step 1 that $\frac{19}{(x - 4)(x^2 + 3)}$

$$\equiv \frac{A(x^2 + 3) + (Bx + C)(x - 4)}{(x - 4)(x^2 + 3)}$$

Step 3: Equate the numerators from step 2:

$\Rightarrow 19 \equiv A(x^2 + 3) + (Bx + C)(x - 4)$

Step 4: Choose suitable x-values to evaluate A, B and C.

From our previous method we know that a suitable x-value is when $(x - 4) = 0 \Rightarrow x = 4$. Substitute this value into the identity in step 3:

$19 = 19A + (Bx + C)(0) \Rightarrow A = 1$

There are no more obvious x-values which we can find by this method but since

$19 \equiv A(x^2 + 3) + (Bx + C)(x - 4)$

is an identity it is true for all values of x so we can choose any values of x to help us evaluate B and C.

Method notes

The values of B and C can be found by choosing any values of x except $x = 4$ which has already been used.

Step 5: Substitute $x = 0$ into the identity in step 3:

$19 = 3A - 4C = 3 - 4C$

but $A = 1$ so $C = -4$

Step 6: Substitute another value for x say $x = 1$ into the identity in step 3:

$19 = 4A + (B + C)(-3)$

but $A = 1$ and $C = -4$ therefore

$19 = 4 - 3B + 12$ so $B = -1$

Step 7: Rewrite the original fraction using the values of A, B and C

hence $\dfrac{19}{(x-4)(x^2+3)} \equiv \dfrac{1}{(x-4)} - \dfrac{(x+4)}{(x^2+3)}$

Method notes

When choosing other x-values let $x = 0$ as this will usually simplify the working.

Stop and think 1

Find the value of the positive integers A, B and C so that $\dfrac{31}{48} = \dfrac{A}{3} + \dfrac{B}{4} + \dfrac{C}{16}$

Fractions with repeated factors in the denominator

Example

Write $\dfrac{x}{(x-1)(x-2)^2}$ in terms of partial fractions.

Answer

Step 1: Let $\dfrac{x}{(x-1)(x-2)^2} \equiv \dfrac{A}{(x-1)} + \dfrac{B}{(x-2)} + \dfrac{C}{(x-2)^2}$

where A, B and C are constants.

Step 2: Rewrite $\dfrac{A}{(x-1)} + \dfrac{B}{(x-2)} + \dfrac{C}{(x-2)^2}$ with their lowest common denominator $(x-1)(x-2)^2$:

$$\frac{A}{(x-1)} + \frac{B}{(x-2)} + \frac{C}{(x-2)^2}$$

$$\equiv \frac{A(x-2)^2}{(x-1)(x-2)^2} + \frac{B(x-1)(x-2)}{(x-1)(x-2)^2} + \frac{C(x-1)}{(x-1)(x-2)^2}$$

From step 1:

$$\frac{x}{(x-1)(x-2)^2} \equiv \frac{A(x-2)^2 + B(x-2)(x-1) + C(x-1)}{(x-1)(x-2)^2}$$

Method notes

This fraction has $(x-2)^2$ as a denominator factor. This type of factor is called a **repeated factor**. The other denominator factor is $(x-1)$. In this case add together the powers of the denominator factors i.e. $1 + 2 = 3$ which means that we need to use three partial fractions. The denominator factors of the partial fractions are $(x-1)$, $(x-2)$ and $(x-2)^2$. Their lowest common denominator is $(x-1)(x-2)^2$ as it is the algebraic expression of the lowest degree which has $(x-1)$, $(x-2)$ and $(x-2)^2$ as factors.

Continued on the next page

Step 3: Equate the numerators from step 2

$$\Rightarrow x \equiv A(x-2)^2 + B(x-2)(x-1) + C(x-1)$$

Step 4: Choose suitable x-values to substitute into the identity in step 3 to find the values of A, B and C.

From our previous method a suitable x-value is when $(x-1)=0$

so $x=1$ in step 3 $\Rightarrow 1 = A + (0)B + (0)C \qquad \Rightarrow A = 1$

A second suitable value is when $(x-2)=0$

so $x=2$ in step 3 $\Rightarrow 2 = (0)A + (0)B + C \qquad \Rightarrow C = 2$

For a third suitable value let $x=0$

So $x=0$ in step 3 $\Rightarrow 0 = 4A + 2B - C \qquad \Rightarrow B = -1$

Step 5: Rewrite the original fraction using the values of A, B and C

hence $\dfrac{x}{(x-1)(x-2)^2} \equiv \dfrac{1}{(x-1)} - \dfrac{1}{(x-2)} + \dfrac{2}{(x-2)^2}$

Method notes

An alternative method for finding the value of B is to compare the coefficients of x^2 in $x \equiv A(x-2)^2 + B(x-2)(x-1) + C(x-1)$ giving $0 = A + B$.

Then $B = -A = -1$

Improper fractions

The examples discussed previously involved proper fractions in which the degree of the numerator was less than the degree of the denominator.

$\dfrac{1}{(x-1)}$, $\dfrac{2}{(x-2)^2}$ and $\dfrac{x}{(x^2-5)}$ are examples of proper fractions.

When the degree of the numerator is equal to or greater than the degree of the denominator, the fraction is called an **improper** fraction.

$\dfrac{x^2+3}{x^2-2}$ and $\dfrac{x^3+2x-3}{x^2-3+4}$ are examples of improper fractions.

Improper fractions are split into partial fractions by first dividing the numerator by the denominator. This means that the improper fraction can then be rewritten in terms of a proper fraction as the following example illustrates.

Example

Express $\dfrac{x^3-x+3}{x^2+x-2}$ in terms of partial fractions.

Answer

Step 1: $\dfrac{x^3 - x + 3}{x^2 + x - 2}$ is an improper fraction so divide the numerator by the denominator:

$$\begin{array}{r|l} & x - 1 \\ x^2 + x - 2 & x^3 - x + 3 \\ & \underline{x^3 + x^2 - 2x} \\ & -x^2 + x + 3 \\ & \underline{-x^2 - x + 2} \\ & 2x + 1 \end{array}$$

Step 2: Rewrite $\dfrac{x^3 - x + 3}{x^2 + x - 2} \equiv (x - 1) + \dfrac{2x + 1}{(x^2 + x - 2)}$

Step 3: Rewrite the proper fraction from step 2 in partial fractions:

$$\frac{2x + 1}{(x^2 + x - 2)} \equiv \frac{2x + 1}{(x + 2)(x - 1)} \equiv \frac{A}{(x + 2)} + \frac{B}{(x - 1)}$$

where A and B are constants.

Step 4: Rewrite $\dfrac{A}{(x + 2)} + \dfrac{B}{(x - 1)}$ with their lowest common denominator $(x + 2)(x - 1)$:

$$\frac{A}{(x + 2)} + \frac{B}{(x - 1)} \equiv \frac{A(x - 1)}{(x - 1)(x + 2)} + \frac{B(x + 2)}{(x - 1)(x + 2)}$$

From step 3 $\dfrac{2x + 1}{(x^2 + x - 2)} \equiv \dfrac{A(x - 1)}{(x - 1)(x + 2)} + \dfrac{B(x + 2)}{(x - 1)(x + 2)}$

Step 5: Equate the numerators from step 4:

$2x + 1 \equiv A(x - 1) + B(x + 2)$

Step 6: Choose suitable x-values to substitute into step 5 to find A and B:

$x = 1: 3 = (0)A + 3B \quad \Rightarrow B = 1$

and $x = -2: -3 = (-3)A + (0)B \Rightarrow A = 1$

Step 7: Rewrite the **original fraction** from steps 2 and 3 using the values of A and B

hence $\dfrac{x^3 - x + 3}{x^2 + x - 2} \equiv (x - 1) + \dfrac{1}{(x + 2)} + \dfrac{1}{(x - 1)}$

Method notes

Algebraic long division was covered in Core 2.

Method notes

$\dfrac{(2x + 1)}{(x^2 + x - 2)}$ has a quadratic function in the denominator which can be factorised

$\Rightarrow x^2 + x - 2 = (x + 2)(x - 1)$.

These **two** factors are then used as the denominators of **two partial fractions.**

Essential notes

You will learn how to integrate algebraic fractions in Chapter 5.

Differentiating algebraic fractions

There are two approaches to differentiating algebraic fractions such as

$$y = \frac{2x+1}{(x-1)(x+2)} \text{ and } y = \frac{x^3 - x + 3}{x^2 + x - 2}.$$

In Core 3 you were introduced to the quotient rule which states:

if $y = \frac{u}{v}$ then $\frac{dy}{dx} = \frac{v\frac{du}{dx} - u\frac{dv}{dx}}{v^2}$ where u and v are both functions of x.

Algebraic fractions can be differentiated using the quotient rule.

However, another approach is to rewrite the algebraic fraction using partial fractions and then differentiate each partial fraction separately as the following examples illustrate.

Exam tips

In the examination you are likely to be asked to find the partial fractions first and then differentiate.

Example

Differentiate the following algebraic fractions with respect to x:

a) $y = \frac{2x+1}{(x-1)(x+2)}$

b) $y = \frac{x^3 - x + 3}{x^2 + x - 2}$

Answer

a) $y = \frac{2x+1}{(x-1)(x+2)}$

Step 1: Rewrite y in terms of partial fractions as explained in the previous section, hence

$$y = \frac{2x+1}{(x-1)(x+2)} \equiv \frac{1}{(x-1)} + \frac{1}{(x+2)}$$

Step 2: Rewrite each partial fraction:

$$y = \frac{1}{(x-1)} + \frac{1}{(x+2)}$$
$$\equiv (x-1)^{-1} + (x+2)^{-1}$$

Step 3: Differentiate each partial fraction using the chain rule:

$$\frac{dy}{dx} = (-1)\,(x-1)^{-2}\,(1) + (-1)(\,x+2)^{-2}\,(1)$$

Method notes

The chain rule

$\frac{dy}{dx} = \frac{dy}{du} \times \frac{du}{dx}$ was covered in Core 3.

It is used to differentiate a composite function of x.

Step 4: Simplify and rewrite as algebraic fractions:

$$\frac{dy}{dx} = -\frac{1}{(x-1)^2} - \frac{1}{(x+2)^2}$$

b) $y = \dfrac{x^3 - x + 3}{x^2 + x - 2}$

Step 1: This is an improper fraction so divide out then rewrite in terms of partial fractions as explained in a previous example,

hence $y = (x-1) + \dfrac{1}{(x+2)} + \dfrac{1}{(x-1)}$

Step 2: Rewrite each partial fraction:

$$y = (x-1) + (x+2)^{-1} + (x-1)^{-1}$$

Step 3: Differentiate with respect to x, using the chain rule for each partial fraction:

$$\frac{dy}{dx} = 1 - \frac{1}{(x+2)^2} - \frac{1}{(x-1)^2}$$

In the last two examples we have seen basic partial fractions or partial fractions in the standard linear denominator form $\dfrac{A}{ax+b}$ where A, a and b are constants.

To differentiate such a partial fraction we used the chain rule as follows:

Let $y = \dfrac{A}{ax+b}$ and $u = ax + b \Rightarrow \dfrac{du}{dx} = a$

and $\dfrac{A}{ax+b} = Au^{-1}$

So $\dfrac{dy}{du} = -Au^{-2} = -\dfrac{A}{u^2}$

The chain rule states: $\dfrac{dy}{dx} = \dfrac{dy}{du} \times \dfrac{du}{dx}$

$\Rightarrow \dfrac{dy}{dx} = -\dfrac{A}{u^2} \times a = -\dfrac{aA}{(ax+b)^2}$ since $u = ax + b$

The following example illustrates how to use this general result for differentiating a basic partial fraction.

Example

Find if $y = \dfrac{19}{(x - 4)(x^2 + 3)}$

Answer

Step 1: Rewrite y in terms of partial fractions as illustrated in the example on page 10 hence $y = \dfrac{19}{(x - 4)(x^2 + 3)} = \dfrac{1}{(x - 4)} - \dfrac{(x + 4)}{(x^2 + 3)}$ then differentiate each partial fraction with respect to x.

Step 2: The first partial fraction $\dfrac{1}{(x - 4)}$ is in the standard linear denominator form so apply the chain rule to differentiate which gives $-\dfrac{1}{(x - 4)^2}$

Step 3: The second partial fraction $\dfrac{x + 4}{(x^2 + 3)}$ is not in the standard linear denominator form of $\dfrac{A}{ax + b}$.

$\dfrac{x + 4}{(x^2 + 3)}$ is a quotient of two functions of x so use the quotient rule to differentiate:

let $p = \dfrac{x + 4}{(x^2 + 3)}$ where $u = (x + 4) \Rightarrow \dfrac{du}{dx} = 1$ and let $v = (x^2 + 3)$

$\Rightarrow \dfrac{dv}{dx} = 2x$

The quotient rule states $\dfrac{dp}{dx} = \dfrac{v\left(\dfrac{du}{dx}\right) - u\left(\dfrac{dv}{dx}\right)}{v^2}$

$$\Rightarrow \frac{dp}{dx} = \frac{1(x^2 + 3) - 2x(x + 4)}{(x^2 + 3)^2} = \frac{-x^2 - 8x + 3}{(x^2 + 3)^2}$$

Step 4: Use the results from steps 2 and step 3 to give:

$$\frac{dy}{dx} = -\frac{1}{(x - 4)^2} - \frac{(-x^2 - 8x + 3)}{(x^2 + 3)^2}$$

$$= -\frac{1}{(x - 4)^2} + \frac{(x^2 + 8x - 3)}{(x^2 + 3)^2}$$

Stop and think answers

1. *If* $\frac{31}{48} = \frac{A}{3} + \frac{B}{4} + \frac{C}{16}$

then find a lowest common denominator for the three fractions on the right hand side of the equation which is 48 and rewrite each of them with this denominator so

$$\frac{31}{48} = \frac{16A}{48} + \frac{12B}{48} + \frac{3C}{48}$$

Therefore as all the denominators are equal in this equation then the numerators must balance so

$31 = 16A + 12B + 3C$

If A, B and C are integers then try $A = 1$ *in the equation so*

$31 - 16 = 12B + 3C \Rightarrow 12B + 3C = 15$

so *if* $B = 1$, $12 + 3C = 15 \Rightarrow 3C = 3$ so $C = 1$

Therefore rewriting the original equation using the values of A, B and C

gives $\frac{31}{48} = \frac{1}{3} + \frac{1}{4} + \frac{1}{16}$

Binomial series for any value of n

Expansion of $(ax + b)^n$ for positive integer n

In Core 2 the binomial series was introduced as an expansion of the expression $(1 + x)^n$

The binomial expansion formula is:

$$(1 + x)^n = 1 + nx + \frac{n(n-1)}{2!}x^2 + \frac{n(n-1)(n-2)}{3!}x^3 + \frac{n(n-1)(n-2)(n-3)}{4!}x^4 + \ldots$$

when **n is a positive integer**. In this case the expansion contains a finite number of terms.

An exact answer can be found when a numerical value of n is substituted into the formula.

Essential notes

The binomial expansion formula will have $(n + 1)$ terms when n is a positive integer.

Essential notes

In this example $n = 5$ so there will be $n + 1 = 6$ terms. The formula will end with the term in x^5

Example

Use the binomial expansion to find $(1 + 2x)^5$.

Answer

Step 1: Compare $(1 + 2x)^5$ with $(1 + x)^n \Rightarrow$ to use the formula result replace 'x' by $2x$ and use $n = 5$

Step 2: Use the binomial expansion formula with 'x' $= 2x$ and $n = 5$

$$(1 + 2x)^5 = 1 + 5 \times (2x) + \frac{5 \times 4}{2!} \times (2x)^2 + \frac{5 \times 4 \times 3}{3!} \times (2x)^3 + \frac{5 \times 4 \times 3 \times 2}{4!} \times (2x)^4 + \frac{5 \times 4 \times 3 \times 2 \times 1}{5!} \times (2x)^5$$

Step 3: Simplify each term:

$$(1 + 2x)^5 = 1 + 10x + \frac{20}{2} \times 4x^2 + \frac{5 \times 4 \times 3}{6} \times 8x^3 + \frac{5 \times 4 \times 3 \times 2}{24} \times 16x^4 + \frac{5 \times 4 \times 3 \times 2 \times 1}{120} \times 32x^5$$

hence $(1 + 2x)^5 = 1 + 10x + 40x^2 + 80x^3 + 80x^4 + 32x^5$

Essential notes

The factorial symbol ! is a shorthand notation so

$4! = 4 \times 3 \times 2 \times 1 = 24$

$6! = 6 \times 5 \times 4 \times 3 \times 2 \times 1 = 720$

In algebra $n! = n \times (n - 1) \times (n - 2) \times (n - 3) \times \ldots \times 1$

Natural numbers are non-negative integers.

The binomial series for fractional and negative indices

The terms and coefficients in the binomial series in the expansion above are simplifications of the expressions ${}^nC_r = \binom{n}{r} = \frac{n!}{r!(n-r)!}$ where n and r are natural numbers. This notation was explained in Core 2

For fractional and negative indices, ${}^nC_r$ has no meaning since the definition ${}^nC_r = \binom{n}{r} = \frac{n!}{r!(n-r)!}$ only applies when n and r are non-negative integers.

However, ${}^nC_r = \dfrac{n!}{r!(n-r)!}$

$$= \frac{n(n-1)(n-2)(n-3)\ldots(n-r+1)(n-r)(n-r-1)(n-r-2)\ldots 1}{r(r-1)(r-2)(r-3)\ldots 3\times 2\times 1\times(n-r)(n-r-1)(n-r-2)\ldots 1}$$

which simplifies to

$${}^nC_r = \frac{n(n-1)(n-2)(n-3)\ldots(n-r+1)}{r(r-1)(r-2)(r-3)\ldots 3\times 2\times 1}$$

For fractional and negative values of n, $(n-r+1)$ gives a non-zero value for all values of r if r is a natural number. For example,

if $n=\frac{3}{4}$: $n-1=-\frac{1}{4}$ $\quad n-2=-\frac{5}{4}$ $\quad n-3=-\frac{9}{4}$ $\quad n-4=-\frac{13}{4}$ and so on

if $n=-2$: $n-1=-3$ $\quad n-2=-4$ $\quad n-3=-5$ and so on

Thus the binomial expansion of $(1+x)^n$ when n is fractional or negative and r is a natural number produces an infinite series as follows:

$$(1+x)^n = 1+nx+\frac{n(n-1)}{2!}x^2+\ldots+\frac{n(n-1)(n-2)\ldots(n-r+1)}{r!}x^r+\ldots$$

Example

Write down the **first five terms** of the binomial expansion of $(1+x)^{\frac{1}{2}}$

Answer

Step 1: Compare $(1+x)^{\frac{1}{2}}$ with $(1+x)^n \Rightarrow$ to use the formula result

replace n by $\frac{1}{2}$

Step 2: Use the binomial expansion formula with $n=\frac{1}{2}$ and stop at the term in x^4:

$$(1+x)^{\frac{1}{2}} = 1+\frac{1}{2}x+\frac{\left(\frac{1}{2}\right)\left(\frac{1}{2}-1\right)}{2!}x^2+\frac{\left(\frac{1}{2}\right)\left(\frac{1}{2}-1\right)\left(\frac{1}{2}-2\right)}{3!}x^3$$

$$+\frac{\left(\frac{1}{2}\right)\left(\frac{1}{2}-1\right)\left(\frac{1}{2}-2\right)\left(\frac{1}{2}-3\right)}{4!}x^4$$

Continued on the next page

Step 3: Simplify each of the five terms

$$(1+x)^{\frac{1}{2}} = 1 + \frac{1}{2}x + \frac{\frac{1}{2}\times\left(-\frac{1}{2}\right)}{2}x^2 + \frac{\frac{1}{2}\times\left(-\frac{1}{2}\right)\times\left(-\frac{3}{2}\right)}{3\times 2}x^3$$

$$+\frac{\left(\frac{1}{2}\right)\left(-\frac{1}{2}\right)\left(-\frac{3}{2}\right)\left(-\frac{5}{2}\right)}{4\times 3\times 2}x^4$$

hence $(1+x)^{\frac{1}{2}} = 1 + \frac{1}{2}x - \frac{1}{8}x^2 + \frac{1}{16}x^3 - \frac{5}{128}x^4$

Exam tips

Only give the number of terms asked for in the question even though the expansion will have an infinite number of terms.

Example

Write down the **first three terms** in the binomial expansion of $\frac{1}{(1-3x)}$

Answer

Step 1: Rewrite $\frac{1}{(1-3x)}$ as $(1-3x)^{-1}$ using a rule of indices.

Step 2: Rewrite $(1-3x)^{-1}$ as $(1+(-3x))^{-1}$ then compare with $(1+x)^n$

To use the formula replace 'x' by $(-3x)$ and n by (-1).

Step 3: Use the binomial expansion formula with 'x' $=(-3x)$ and $n=(-1)$ and stop at the term in x^2:

$$\frac{1}{(1-3x)} = (1+(-3x))^{-1} = 1 + (-1)\times(-3x) + \frac{(-1)(-2)}{2!}(-3x)^2$$

Step 4: Simplify each of the first three terms

hence $\frac{1}{(1-3x)} = 1 + 3x + 9x^2$

Method notes

The binomial expansion formula has a plus (+) sign between the two terms inside the bracket. When there is a minus (–) sign between the two terms in the bracket then you must rewrite the bracket $(1-3x) = (1+(-3x))$ before you use the formula.

Stop and think 1

a) Write down the expansion of $\frac{1}{(1-3x)}$ *as a four term series.*

b) Write down the expansion of $\frac{1}{(1-3x)}$ *as a five term series.*

c) If $x = 0.2$ *does the binomial series for* $\frac{1}{(1-3x)}$ *converge or diverge?*

d) If $x = 0.4$ *does the binomial series for* $\frac{1}{(1-3x)}$ *converge or diverge?*

Give reasons for your answers.

Convergence of the binomial series

We now consider the first six terms in the binomial expansion of $\frac{1}{(1+2x)}$.

$\frac{1}{(1+2x)} = (1+2x)^{-1}$ so we use 'x' $= 2x$ and $n = (-1)$ and stop at the term in x^5 in the binomial expansion formula therefore

$$\frac{1}{(1+2x)} = 1 + (-1)(2x) + \frac{(-1)(-2)(2x)^2}{2} \ldots = 1 - 2x + 4x^2 - 8x^3 + 16x^4 - 32x^5$$

The following table shows the values of the binomial expansion for different lengths of this series and for different values of x:

x	Two-term series	Three-term series	Four-term series	Five-term series	Six-term series
–1.0	3	7	15	31	63
–0.6	2.2	3.64	5.368	7.4416	9.92992
–0.4	1.8	2.44	2.952	3.3616	3.68928
–0.1	1.2	1.24	1.248	1.2496	1.24992
0.1	0.8	0.84	0.832	0.8336	0.83328
0.4	0.2	0.84	0.328	0.7376	0.40992
0.6	–0.2	1.24	–0.488	1.5856	–0.90272
1.0	–1	3	–5	11	–21

If we work out the exact values of $\frac{1}{(1+2x)}$ for the given x–values they are:

for $x = 0.1$ $\frac{1}{(1+2x)} = 0.833333$ for $x = -0.1$ $\frac{1}{(1+2x)} = 1.25$

for $x = 0.4$ $\frac{1}{(1+2x)} = 0.555556$ for $x = -0.4$ $\frac{1}{(1+2x)} = 5$

for $x = 0.6$ $\frac{1}{(1+2x)} = 0.454546$ for $x = -0.6$ $\frac{1}{(1+2x)} = -5$

for $x = 1.0$ $\frac{1}{(1+2x)} = 0.333333$ for $x = -1.0$ $\frac{1}{(1+2x)} = -1$

From this you can see that for $x = -0.4$, $x = -0.1$, $x = 0.1$ and $x = 0.4$ in the table results, the binomial series appears to be getting closer and closer to the exact value so we can say that the series is **converging**.

For $x = -1.0$, $x = -0.6$, $x = 0.6$ and $x = 1.0$ in the table results, the binomial series appears to be getting further away for the exact value so we say that the series is **diverging**.

If we continued this process with more values of x we would find that if we take x in the range $-0.5 < x < 0.5$ then the binomial series converges but outside this range it diverges.

Essential notes

Converging and diverging series were explained in Core 2

The modulus notation $|x|$, the absolute or numerical value of x, was explained in Core 3

So $-\frac{1}{2} < x < \frac{1}{2} \Rightarrow |x| < \frac{1}{2}$

Hence the critical values for convergence of the binomial expansion of $\frac{1}{(1+2x)}$ is the range $-0.5 < x < 0.5$ or $-\frac{1}{2} < x < \frac{1}{2}$ or $|x| < \frac{1}{2}$.

Condition for convergence of the binomial series

In the last example, the expansion of $\frac{1}{(1+2x)} = (1+2x)^{-1}$

we found the series converges if $|x| < \frac{1}{2} \Rightarrow |2x| < 1$

If we had started with the expansion of $(1+3x)^{-1}$ we would have found the series converges for $|x| < \frac{1}{3} \Rightarrow |3x| < 1$

If we had started with the expansion of $(1-5x)^{-2} = (1+(-5x))^{-2}$ we would have found the series converges for $|-5x| < 1 \Rightarrow |5x| < 1$

Hence in general terms:

the binomial expansion of $(1+ax)^n$ for any fractional or negative value of n, converges provided $-1 < ax < 1$ or $|ax| < 1$ or $|x| < \frac{1}{a}$ where a is a constant.

Exam tips

Learn how to find the condition of convergence for a binomial series.

Example

a) Find the first three terms in the binomial expansion of $(1-5x)^{\frac{3}{4}}$.

b) State the range of values of x for which the series converges.

Answer

a) **Step 1:** Rewrite $(1-5x)^{\frac{3}{4}}$ as $(1+(-5x))^{\frac{3}{4}}$ and compare this with $(1+x)^n$ to use the binomial expansion formula 'x' = $(5x)$ and $n = \frac{3}{4}$

Step 2: Use the binomial expansion formula with 'x' = $(5x)$ and $n = \frac{3}{4}$ and stop at the term in x^2:

$$\Rightarrow (1-5x)^{\frac{3}{4}} = 1 + \left(\frac{3}{4}\right) \times (-5x) + \frac{\left(\frac{3}{4}\right)\left(\frac{3}{4}-1\right)}{2} \times (-5x)^2$$

Step 3: Simplify each of the three terms:

$$(1-5x)^{\frac{3}{4}} = 1 - \frac{15}{4}x - \frac{75}{32}x^2$$

b) The test for convergence of the binomial series $(1+ax)^n$ is that $|ax| < 1$ and in this question $(1-5x)^{\frac{3}{4}}$, $a = -5$ therefore the series converges for $|-5x| < 1$

There is convergence if $|5x| < 1$ so $-\frac{1}{5} < x < \frac{1}{5}$

Use of the binomial series for approximations

The binomial series is often useful for finding approximate values of functions as the following example illustrates.

Example

a) Find a quadratic function which approximates to $\sqrt{(9+x)}$ for small values of x.

b) Use the function to find the approximate value of $\sqrt{8.97}$ to 4 s.f.

Answer

a) You should recognise from the wording of the question that because a quadratic function is asked for it means that the binomial expansion must be used up to and including the term in x^2

Step 1: Rewrite $\sqrt{(9+x)}$ as $(9+x)^{\frac{1}{2}}$ and rewrite the algebra inside the bracket so that it can be compared with $(1+ax)^n$.

Step 2: Rewrite the algebra $(9+x)=9\left(1+\frac{x}{9}\right)$ so that it can be compared with $(1+ax)$

$$\Rightarrow (9+x)^{\frac{1}{2}}=\left(9\left(1+\frac{x}{9}\right)\right)^{\frac{1}{2}}=(9)^{\frac{1}{2}}\left(1+\frac{x}{9}\right)^{\frac{1}{2}} \text{ using rules of indices.}$$

Step 3: Simplify the algebra:

$$(9+x)^{\frac{1}{2}}=\sqrt{(9)}\times\sqrt{\left(1+\frac{x}{9}\right)}=3\left(1+\frac{x}{9}\right)^{\frac{1}{2}}$$

Step 4: Use the binomial expansion formula for $\left(1+\frac{x}{9}\right)^{\frac{1}{2}}$ with 'x' $=\frac{x}{9}$ and $n=\frac{1}{2}$ and stop at the term in x^2:

$$\left(1+\frac{x}{9}\right)^{\frac{1}{2}}=1+\left(\frac{1}{2}\right)\times\left(\frac{x}{9}\right)+\frac{\left(\frac{1}{2}\right)\left(\frac{1}{2}-1\right)}{2}\times\left(\frac{x}{9}\right)^2$$

Step 5: Simplify each of the three terms $\left(1+\frac{x}{9}\right)^{\frac{1}{2}}=1+\frac{x}{18}-\frac{x^2}{648}$

Step 6: Substitute the expansion into the result in step 3:

$$\Rightarrow \sqrt{(9+x)}=3\left(1+\frac{x}{9}\right)^{\frac{1}{2}}\approx 3\left(1+\frac{x}{18}-\frac{x^2}{648}\right)\approx 3+\frac{x}{6}-\frac{x^2}{216}$$

This series will converge for $|x|<9$

Continued on the next page

Method notes

The binomial expansion gives 2.9949958 (to 8 s.f.)

A calculator gives

$\sqrt{8.97} = 2.9949958$

The binomial expansion is in fact correct to 8 s.f. in this case.

b) If you are to use the quadratic function from part (a) which approximated to $\sqrt{(9+x)}$, it means if you are to approximate $\sqrt{8.97}$ you must compare $\sqrt{(9+x)}$ and $\sqrt{8.97}$

Step 1: If $\sqrt{(9+x)} = \sqrt{8.97}$ then $x = -0.03$

Step 2: Let $x = -0.03$ in the result from step 6 of part (a)

$$\Rightarrow \sqrt{(9+x)} = \sqrt{(9+(-0.03))} = \sqrt{8.97}$$

$$\approx 3 + \frac{-0.03}{6} - \frac{(-0.03)^2}{216} = 2.995 \text{ (4 s.f.)}$$

Using partial fractions with the binomial series

You may be asked to approximate more complicated algebraic fractions, such as $\frac{5x+3}{(2x-3)(x+2)}$ by using a binomial series expansion. In this case it is often easier to use partial fractions to simplify the process as the following example illustrates.

Example

a) Write $\frac{5x+3}{(2x-3)(x+2)}$ as partial fractions.

b) Find a quadratic approximation of $\frac{5x+3}{(2x-3)(x+2)}$.

c) State the range of values of x for which the binomial expansion is valid.

Method notes

Partial fractions were explained in Chapter 1

This is a proper fraction with two linear factors in the denominator which means we can write this as a sum of two partial fractions, using the two denominator factors.

Answer

a) **Step 1:** Let $\frac{5x+3}{(2x-3)(x+2)} \equiv \frac{A}{2x-3} + \frac{B}{x+2}$ where A and B are constants.

Step 2: Write the two partial fractions with a common denominator:

$$\frac{5x+3}{(2x-3)(x+2)} \equiv \frac{A(x+2) + B(2x-3)}{(2x-3)(x+2)}$$

Step 3: Equate the numerators of the fractions in step 2:

$$5x + 3 \equiv A(x+2) + B(2x-3)$$

Step 4: Choose suitable values of x to evaluate A and B from step 4:

$$x = -2 \Rightarrow -7 = (0)A + (-7)B \Rightarrow B = 1$$

$$x = \frac{3}{2} \Rightarrow \frac{21}{2} = \left(\frac{7}{2}\right)A + (0)B \Rightarrow A = 3$$

Step 5: Rewrite the original fraction using the values of A and B

Hence $\dfrac{5x+3}{(2x-3)(x+2)} \equiv \dfrac{3}{2x-3} + \dfrac{1}{x+2}$

b) For the quadratic approximation, rewrite each partial fraction so that the binomial expansion formula can be used up to and including the term in x^2.

Step 1: Rewrite $\dfrac{3}{(2x-3)} = \dfrac{3}{(3-2x)} = \dfrac{3}{3\left(1-\frac{2}{3}x\right)}$

$$= \frac{1}{\left(1-\frac{2}{3}x\right)} = \left(1-\frac{2}{3}x\right)^{-1}$$

$$= \left(-\frac{1}{3}(3-2x)\right)^{-1} = -\left(1-\frac{2}{3}x\right)^{-1} \text{ and compare with } (1+x)^n$$

$$\Rightarrow \text{'}x\text{'} = \left(-\frac{2}{3}x\right) \text{ and } n = -1$$

Step 2: Use the binomial expansion formula with $'x' = \left(-\frac{2}{3}x\right)$ and $n = -1$

$$\Rightarrow \frac{3}{2x-3} \approx -\left(1+\frac{2}{3}x+\left(\frac{2}{3}x\right)^2\right) = -1-\frac{2}{3}x-\frac{4}{9}x^2$$

Step 3: Rewrite $\dfrac{1}{(x+2)} = \dfrac{1}{(2+x)} = \dfrac{1}{2\left(1+\frac{1}{2}x\right)} = \dfrac{1}{2}\left(1+\dfrac{1}{2}x\right)^{-1}$

$$= \frac{1}{2}\left(1+\frac{x}{2}\right)^{-1} \text{ and compare with } (1+x)^n \Rightarrow \text{'}x\text{'} = \left(\frac{x}{2}\right) \text{ and } n = -1$$

Step 4: Use the binomial expansion formula with $'x' = \left(\frac{x}{2}\right)$ and $n = -1$

$$\approx \frac{1}{2}\left(1-\frac{x}{2}+\left(\frac{x}{2}\right)^2\right) = \frac{1}{2}-\frac{x}{4}+\frac{x^2}{8}$$

Step 5: Use the results from steps 2 and 4 so:

$$\frac{5x+3}{(2x-3)(x+2)} \approx \left(-1-\frac{2}{3}x-\frac{4}{9}x^2\right)+\left(\frac{1}{2}-\frac{x}{4}+\frac{x^2}{8}\right)$$

Step 6: Collect together like terms:

$\dfrac{5x+3}{(2x-3)(x+2)} \approx -\dfrac{1}{2}-\dfrac{11x}{12}-\dfrac{23x^2}{72}$ which gives the quadratic approximation.

Method notes

Write the denominator of the algebraic fraction in the form $(1+ax)^n$

Continued on the next page

c) The binomial expansion of $(1 + ax)^n$ is valid for any fractional or negative value of n

if $-1 < ax < 1$ or $|ax| < 1$ or $|x| < \frac{1}{a}$ where a is a constant.

In this example there were two binomial series expansions and for their convergence (or validity) the range of values of x must satisfy both expansions.

Step 1: Find the validity range for x in the expansion of $-\left(1 - \frac{2}{3}x\right)^{-1}$

$$|x| < \frac{1}{a} \text{ and in this case } a = -\frac{2}{3} \Rightarrow |x| < \frac{1}{a} \text{ so } |x| < \frac{3}{2}$$

Step 2: Find the validity range for x in the expansion of $\frac{1}{2}\left(1 + \frac{x}{2}\right)^{-1}$

$$|x| < \frac{1}{a} \text{ and in this case } a = \frac{1}{2} \text{ so } |x| < 2$$

Step 3: Find the validity range for both expansions:

$|x| < \frac{3}{2}$ and $|x| < 2$ must both be satisfied

therefore $|x| < \frac{3}{2}$

hence the quadratic expansion of $\frac{5x + 3}{(2x - 3(x + 2)}$

is valid for $|x| < \frac{3}{2}$

Stop and think answers

a) *Using the binomial expansion formula and rewriting* $1 - 3x$ *as* $1 + (-3x)$ *then*

$$(1-3x)^{-1} = 1 + (-1)\,(-3x) + \frac{(-1)(-2)(-3x)^2}{2 \times 1} + \frac{(-1)(-2)(-3)(-3x)^3}{3 \times 2 \times 1} + \ldots$$

$$= 1 + 3x + 9x^2 + 27x^3 + \ldots$$

b) *The next term to give a five term series will be*

$$\frac{(-1)(-2)(-3)(-4)(-3x)^4}{4 \times 3 \times 2 \times 1} = 81x^4$$

so the five terms are $= 1 + 3x + 9x^2 + 27x^3 + 81x^4 ..$

c) *If* $x = 0.2$ *using the answer from part (b)*

then $(1-3x)^{-1} = 1 + 3(0.2) + 9(0.2)^2 + 27(0.2)^3 + 81(0.2)^4 ..$

$$= 1 + 0.6 + 0.36 + 0.216 + 0.1296 \; ..$$

and we can see that the series is converging. The condition for convergence is $|3x| < 1$ *and in this question* $x = 0.2$ *which satisfies this convergence condition.*

If $x = 0.4$ *then* $(1-3x)^{-1} = 1 + 3(0.4) + 9(0.4)^2 + 27(0.4)^3 + 81(0.4)^4 ..$

$$= 1 + 1.2 + 1.44 + 1.728 + 2.0736 \; \ldots$$

and we can see that the series is diverging as the terms are getting larger and also because $|1.2| > 1$.

Parametric equations of curves

In previous modules the equations of curves have always been given in terms of Cartesian coordinates, x and y. You have seen that $y = 3x^2 - 2x - 4$ is the equation of a parabola and $(x - 1)^2 + (y + 2)^2 = 16$ is the equation of the circle with centre $(1, -2)$ and radius 4

Sometimes it is more convenient to express x and y in terms of a third quantity (or variable) called a **parameter**. This means that the curve will be described by a pair of equations, the first equation giving x in terms of the parameter and the second equation giving y in terms of the parameter. You can choose any letter to represent the third variable (or parameter) and often we choose the letter t.

Essential notes

If x is given in terms of another variable (t) it means that x is a function of t and we write $x = f(t)$. The function notation was covered in Core 1

Therefore the pair of equations $x = f(t)$ and $y = g(t)$ are called the **parametric equations** of the curve in terms of a parameter t.

The curve is then said to be given in parametric form.

Example

Sketch the graph given by the parametric equations $x = t + 2$ and $y = 2t - 3$

Answer

Step 1: Choose various values of t and work out the corresponding values of x and y from the given parametric equations.

$\Rightarrow$ if $t = -4$, $x = (-4 + 2) = -2$ and $y = 2(-4) - 3 = -11$ and so on.

Step 2: Complete the table of values for t, x and y.

t	−4	−2	0	2	4
$x = t + 2$	−2	0	2	4	6
$y = 2t - 3$	−11	−7	−3	1	5

Essential notes

Cartesian axes have x as the horizontal axis and y as the vertical axis.

Step 3: Plot the values of x and y in the usual way, using Cartesian axes.

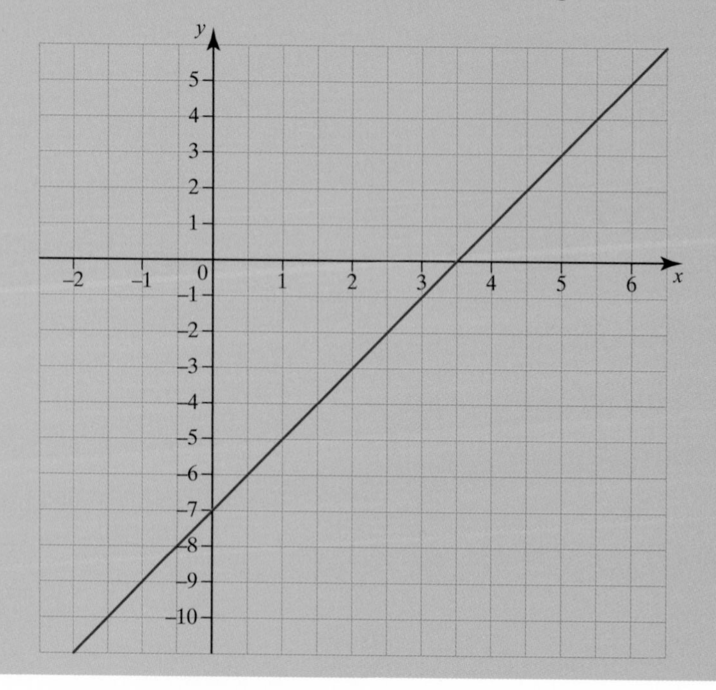

Fig. 3.1
Graph of the line with parametric equations $x = t + 2$ and $y = 2t - 3$

From Figure 3.1 we see that when the points are joined the graph is a straight line.

Example

Sketch the curve given by the parametric equations $x = t + 1$ and $y = t^2$.

Answer

Step 1: Complete a table of values for t, x and y by choosing various values of t to work out the corresponding values of x and y.

t	−3	−2	−1	0	1	2
$x = t + 1$	−2	−1	0	1	2	3
$y = t^2$	9	4	1	0	1	4

Step 2: Plot the values of x and y in the usual way, using Cartesian axes.

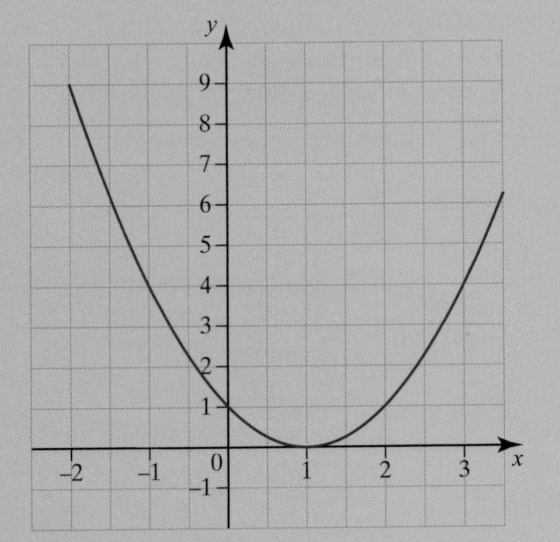

Fig. 3.2
Graph of the curve with parametric equations $x = t + 1$ and $y = t^2$.

From this we can see that when the points are joined the graph is a parabola.

Parametric equations and properties of curves

The parametric equations of a curve can also be used to describe the properties of the curve as the following examples illustrate.

Example

A curve is defined parametrically by the equations

$x = 2t - 1$ and $y = t^3 + 8$

The curve passes through the points $(a, 0)$ and $(0, b)$ where a and b are constants.

Find the values of a and b.

Continued on the next pages

Answer

As the curve passes through the point with Cartesian coordinates $(a, 0)$ this means that $x = a$ and $y = 0$ must satisfy the parametric equations of this curve.

Step 1: Substitute $y = 0$ into the parametric equation $y = t^3 + 8$ and find t:

$\Rightarrow 0 = t^3 + 8$

$\Rightarrow t = -2$

Step 2: Substitute $t = -2$ into the parametric equation $x = 2t - 1$

$\Rightarrow x = 2 \times (-2) - 1 = -5$

So $x = -5$ when $y = 0$ but we know that $x = a$ when $y = 0$ is a point on the curve
therefore $a = -5$

The curve also passes through the point with Cartesian coordinates $(0, b)$ which means that $x = 0$ when $y = b$ must satisfy the parametric equations of this curve.

Step 3: Substitute $x = 0$ into the parametric equation $x = 2t - 1$

$\Rightarrow 0 = 2t - 1 \Rightarrow t = \frac{1}{2}$

Step 4: Substitute $t = \frac{1}{2}$ into the parametric equation $y = t^3 + 8$

$$\Rightarrow y = \left(\frac{1}{2}\right)^3 + 8 = \frac{1}{8} + 8 = \frac{65}{8}$$

but we know that when $x = 0$ $y = b$ is a point on the curve

therefore $b = \frac{65}{8}$

Example

The straight line $2y = 3x - 11$ cuts the curve with parametric equations $x = t^2 + 3$ and $y = 3t - 1$ at the points A and B. Calculate the coordinates of A and B.

Answer

Step 1: Substitute the x and y coordinates of the curve,

$x = t^2 + 3$ and $y = 3t - 1$ into the equation of the straight line

$2y = 3x - 11$ and find the t values:

$\Rightarrow 2(3t - 1) = 3(t^2 + 3) - 11$

$\Rightarrow \quad 6t - 2 = 3t^2 - 2$

$\Rightarrow \quad 3t^2 - 6t = 0$

$\Rightarrow 3t(t-2)=0$

therefore $t=0$ or $t=2$

Step 2: Substitute $t=0$ into the parametric equations $x=t^2+3$ and $y=3t-1$ to find the x and y coordinates of the intersection point:

When $t=0$, $x=t^2+3$

$\Rightarrow \quad x=(0)^2+3 \Rightarrow x=3$

When $t=0$, $y=3t-1$

$\Rightarrow \quad y=3(0)-1 \Rightarrow y=-1$

Step 3: Repeat the process of step 2 with $t=2$

$\Rightarrow x=7$ and $y=5$

Hence the coordinates of the points of intersection of the line and the curve are (3, −1) and (7, 5).

Method notes

The intersection of lines and curves was covered in Core 1

At the points of intersection the x and y coordinates of the curve must also satisfy the equation of the line.

Conversion from parametric form to Cartesian form

In the first example of this chapter, the graph given by the parametric equations $x=t+2$ and $y=2t-3$ was a straight line. From the table of values in the answer, we can deduce that two points on the straight line have coordinates (0, −7) and (4, 1). These are therefore the Cartesian coordinates of two points on the line.

Let A be the point (0, −7) and B be the point (4, 1).

Using the formula for the equation of a line joining two points with $x_1=0$ and $y_1=-7$, $x_2=4$ and $y_2=1$ gives:

$$\frac{y-(-7)}{x}=\frac{1-(-7)}{4-0}=2 \Rightarrow y=2x-7$$

This is the familiar equation of a straight line in Cartesian coordinates, often called the **Cartesian form** of the graph.

Essential notes

The equation of line joining two points $A(x_1, y_1)$ and $B(x_2, y_2)$ is given by:

$$\frac{(y-y_1)}{(x-x_1)}=\frac{(y_2-y_1)}{(x_2-x_1)}$$

This was covered in Core 1

However, it is not necessary to work out the table of values for x and y from the parametric equations in order to obtain the Cartesian form of the graph.

An alternative approach is as follows.

The parametric equations are $x=t+2 \quad (1)$

$y=2t-3 \quad (2)$

These are simultaneous equations so from (1)

$x=t+2 \Rightarrow t=x-2 \quad (3)$

Substituting for t from equation (3) into equation (2)

$\Rightarrow y=2(x-2)-3=2x-7$

$\Rightarrow y=2x-7$

Method notes

The labelling and solution of simultaneous equations was covered in Core 1

Hence the Cartesian equation of the line is $y = 2x - 7$

Eliminating the parameter, t, between the given parametric equations of a graph, is the usual method of approach for finding the Cartesian equation of that graph.

Example

Find the Cartesian equation of the curve represented in parametric form by $x = 5t + 1$ and $y = \frac{2}{t}$

Answer

Step 1: Label the parametric equations $x = 5t + 1$ (1)

and $y = \frac{2}{t}$ (2)

Step 2: Rewrite equation (1) to find $t \Rightarrow t = \frac{x - 1}{5}$ (3)

Step 3: Substitute the t-expression from equation (3) into equation (2):

$$y = \frac{2}{t} \Rightarrow y = \frac{2}{\left(\frac{x-1}{5}\right)} = \frac{10}{x - 1}$$

Hence the Cartesian equation of the curve is $y = \frac{10}{(x - 1)}$

In the two previous examples, the equations given in parametric form have not involved trigonometric ratios. We now consider how to find the Cartesian equation of a curve when the parametric equations do involve trigonometric ratios.

Essential notes

Any letter can be used to represent a parameter. When the parametric equations involve trigonometric ratios, θ is often used as the parameter.

Example

Find the Cartesian form of the equation of a curve which is given in parametric form by $x = 2\cos\theta + 3$ and $y = 2\sin\theta - 1$ where θ is the parameter.

Answer

Step 1: Rewrite the two parametric equations making the trigonometric ratios the subject of each equation:

if $x = 2\cos\theta + 3$ then $\cos\theta = \frac{x - 3}{2}$

and if $y = 2\sin\theta - 1$ then $\sin\theta = \frac{y + 1}{2}$

Step 2: Using the trigonometric identity $\cos^2\theta + \sin^2\theta \equiv 1$ substitute for $\cos\theta$ and $\sin\theta$ from step 1:

$$\left(\frac{x-3}{2}\right)^2 + \left(\frac{y+1}{2}\right)^2 = 1 \text{ so } (x-3)^2 + (y+1)^2 = 4$$

Hence the Cartesian equation of the curve is $(x-3)^2 + (y+1)^2 = 4$

which is the equation of a circle centre $(3, -1)$ and radius 2

Essential notes

The general equation of a circle centre (a, b) and radius r is $(x-a)^2 + (y-b)^2 = r^2$.

This was covered in Core 2

The parameter θ which was used in the last example has a simple geometric interpretation. Figure 3.3 below shows the circle centre the origin, radius 3 and passing through the point P. This circle has the Cartesian equation $x^2 + y^2 = 9$

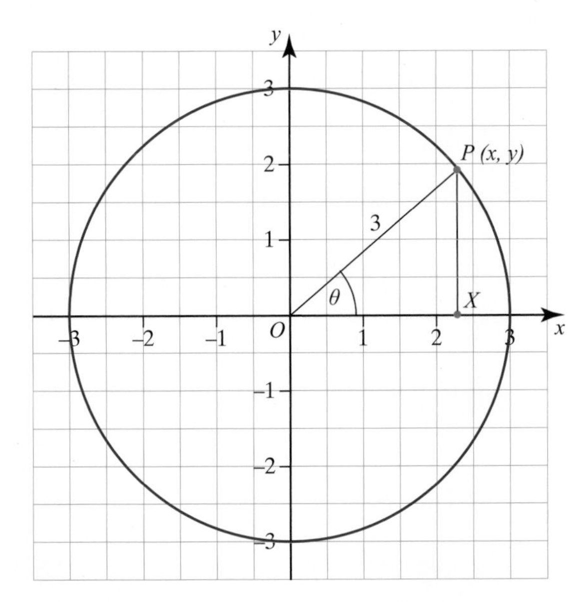

Fig. 3.3
The graph of the circle with equation $x^2 + y^2 = 9$

In Fig. 3.3, the angle between the line OP and the x-axis is denoted by θ.

The line drawn from P, perpendicular to the x-axis, meets the x-axis at the point x so that triangle OPX is a right angled triangle.

In $\Delta\ OPX$: $\dfrac{OX}{OP} = \cos\theta \Rightarrow \dfrac{OX}{3} = \cos\theta$ so $OX = 3\cos\theta$.

Similarly in $\Delta\ OPX$: $\dfrac{PX}{OP} = \sin\theta \Rightarrow \dfrac{PX}{3} = \sin\theta$ so $PX = 3\sin\theta$.

But P has coordinates (x, y) so $OX = x$ and $PX = y$

$\Rightarrow x = 3\cos\theta$ and $y = 3\sin\theta$.

Eliminating θ by using the identity $\cos^2\theta + \sin^2\theta \equiv 1$

and substituting for $\cos\theta$ and $\sin\theta$

$\Rightarrow \left(\dfrac{x}{3}\right)^2 + \left(\dfrac{y}{3}\right)^2 \equiv 1$ so $x^2 + y^2 = 9$ which is the Cartesian equation of the circle centre $(0, 0)$ radius 3

Essential notes

Polar coordinates (r, θ) are part of the Further Mathematics modules.

r is the length of OP and θ is the angle shown.

θ is a parameter called the **polar angle** and is part of a coordinate system called **polar coordinates.**

Parametric differentiation

There are two methods of finding the slope of the tangent to a curve which is defined in terms of parametric equations as the following example illustrates.

Method notes

$\frac{dy}{dx}$ gives the slope of the tangent to a curve at any point on the curve. This was covered in Core 1

Example

A curve is given by the parametric equations $x = 5t$ and $y = 2 - 3t^2$. Find the slope of the tangent to the curve at any point on the curve.

Answer

Method 1 uses the Cartesian equation of the curve to find $\frac{dy}{dx}$.

Step 1: Given $x = 5t$ find $t \Rightarrow t = \frac{x}{5}$.

Step 2: Given $y = 2 - 3t^2$, substitute in the value of 't' from step 1:

$$\Rightarrow y = 2 - 3\left(\frac{x}{5}\right)^2$$ which is the Cartesian equation of the curve.

Step 3: Differentiate with respect to x the Cartesian equation from step 2:

$$\Rightarrow \frac{dy}{dx} = 0 - 3\left(2\left(\frac{x}{5}\right)^1\right)\left(\frac{1}{5}\right) = -\frac{6x}{25}$$

Method notes

$3\left(\frac{x}{5}\right)^2$ is a function of a function of x which means it is differentiated using the chain rule. The chain rule was covered in Core 3

Method 2 uses the parametric equations of the curve and a special case of the chain rule to find $\frac{dy}{dx}$.

Step 1: Given $x = 5t$ differentiate with respect to 't' $\Rightarrow \frac{dy}{dt} = 5$

Step 2: Given $y = 2 = 3t^2$, differentiate with respect to 't' $\Rightarrow \frac{dy}{dt} = -6t$

Step 3: Given $x = 5t \Rightarrow t = \frac{x}{5}$ so substituting for t in step 2 gives

$$\frac{dy}{dt} = -\frac{6x}{5}$$

Step 4: Divide $\frac{dy}{dt}$ by $\frac{dx}{dt} \Rightarrow \frac{\frac{dy}{dt}}{\frac{dx}{dt}} = \frac{-\frac{6x}{5}}{5} = -\frac{6x}{25}$ which was the result we obtained for $\frac{dy}{dx}$ using method 1

Exam tips

Look carefully at the information given in the question before deciding which method to use to find. If elimination of the parameter leads to a very complicated Cartesian equation for the curve then Method 2 is probably the best approach!

The example above shows the relationship between $\frac{dy}{dx}, \frac{dy}{dt}$, and $\frac{dx}{dt}$

which is $\frac{dy}{dx} = \frac{\frac{dy}{dt}}{\frac{dx}{dt}}$.

This can be rewritten as $\frac{dy}{dx} = \frac{dy}{dt} \times \frac{dt}{dx}$ which is a special case of the chain rule for differentiation.

Essential notes

The chain rule for differentiation of a function of a function was covered in Core 3 and you should learn this special case of the chain rule.

Stop and think 1

1. a) *Find the gradient of the normal to the curve given by the parametric equations*

   $x = 5t$ *and* $y = 2 - 3t^2$

   i) *at any point on the curve*

   ii) *at the point* $(5, -1)$.

   b) *Find* $\frac{d^2y}{dt^2}$ *at any point on this curve.*

   c) *Find* $\frac{d^2x}{dt^2}$ *at any point on this curve.*

Example

A curve is defined parametrically by the equations $x = t^2 - 3t$ and $y = t^2 + 3t$

a) Find the slope of the tangent at the point $(-2, 10)$.

b) Find the equation of the tangent at the point $(-2, 10)$.

c) Find the equation of the normal at the point $(-2, 10)$.

Answer

a) **Step 1:** At the point $(-2, 10)$, $x = t^2 - 3t = -2$ (1)

and $y = t^2 + 3t = 10$ (2)

Step 2: Solve these two equations simultaneously:

From equation (1), $t^2 = -2 + 3t$

Step 3: Substitute for t^2 from step 2 into step 1:

$(-2 + 3t) + 3t = 10$ so $6t = 12 \Rightarrow t = 2$

Step 4: Given $y = t^2 + 3t$ find $\frac{dy}{dt}$

$\Rightarrow \frac{dy}{dt} = 2t + 3$

Step 5: Given $x = t^2 - 3t$ find $\frac{dx}{dt}$

$\Rightarrow \frac{dx}{dt} = 2t - 3$

Method notes

Given the Cartesian coordinates of a point on the curve which has been described parametrically, it is useful to find the 't' value at this point. This then gives the connection between x, y, and t at the given point.

Continued on the next page

Step 6: Find $\frac{dy}{dx}$ by using $\frac{dy}{dx} = \frac{\frac{dy}{dt}}{\frac{dx}{dt}}$

so the slope of the tangent at (–2, 10) is the value of $\frac{dy}{dx} = \frac{\frac{dy}{dt}}{\frac{dx}{dt}}$ when $t = 2$

$$\Rightarrow \frac{dy}{dx} = \frac{\frac{dy}{dt}}{\frac{dx}{dt}} = \frac{2t+3}{2t-3}$$

Step 7: Substitute $t = 2$ into $\frac{dy}{dx} = \frac{2t+3}{2t-3}$ so $\frac{dy}{dx} = \frac{7}{1} = 7$ which is the gradient of the tangent to the curve at the point (–2, 10).

b) The equation of the tangent at the point (–2, 10) is given by:

$$\frac{y-10}{x-(-2)} = 7 \Rightarrow y - 10 = 7(x+2)$$

so the required equation is $y = 7x + 24$

c) The gradient of the normal at the point (–2, 10) is $-\frac{1}{7}$

therefore the equation of the normal at the point (–2, 10) is given by:

$$\frac{y-10}{x-(-2)} = -\frac{1}{7} \Rightarrow y - 10 = -\frac{1}{7}(x+2)$$

so the required equation is $y = -\frac{1}{7}x + \frac{68}{7}$

Method notes

b) The equation of the tangent to a curve is given by the formula $\frac{y-y_1}{x-x_1} = m$ where (x_1, y_1) is a point on the tangent and m is the gradient of the tangent at that point. This work was covered in Core 1

The tangent and normal are perpendicular lines so their gradients when multiplied together give –1

This work was covered in Core 1

Method notes

Stationary points of a curve occur when $\frac{dy}{dx} = 0$

Given the parametric equations of the curve,

$x = 2t^3$ and $y = t^2 - 3t + 2$,

Method 2 for finding $\frac{dy}{dx}$ (as explained earlier in the chapter) will be the best approach.

Example

Find and classify the stationary points of the curve with parametric equations $x = 2t^3$ and $y = t^2 - 3t + 2$

Answer

Step 1: Given $x = 2t^3 \Rightarrow \frac{dx}{dt} = 6t^2$ and

given $y = t^2 - 3t + 2 \Rightarrow \frac{dy}{dt} = 2t - 3$

Step 2: $\frac{dy}{dx} = \frac{\frac{dy}{dt}}{\frac{dx}{dt}}$ so substituting into this the results from step 1

$$\Rightarrow \frac{dy}{dx} = \frac{\frac{dy}{dt}}{\frac{dx}{dt}} = \frac{2t - 3}{6t^2}$$

Step 3: Stationary points occur when $\frac{dy}{dx} = 0$ so using the result from step 2 gives $\frac{dy}{dx} = \frac{2t - 3}{6t^2} = 0 \Rightarrow 0 = 2t - 3 \Rightarrow t = \frac{3}{2}$

Step 4: Substitute $t = \frac{3}{2}$ into the parametric equations to find the coordinates of the stationary points:

$$x = 2t^3 \Rightarrow x = 2\left(\frac{3}{2}\right)^3 = \frac{27}{4} \text{ and } y = t^2 - 3t + 2$$

$$\Rightarrow y = \left(\frac{3}{2}\right)^2 - 3\left(\frac{3}{2}\right) + 2 = -\frac{1}{4}$$

As there is only one x value and one y value we know that there is only one stationary point. Therefore the coordinates of the stationary point are $\left(\frac{27}{4}, -\frac{1}{4}\right)$.

$\frac{d^2y}{dx^2} = \frac{\frac{d}{dt}\left(\frac{dy}{dx}\right)}{\frac{dx}{dt}}$ and we know from step 2 that $\frac{dy}{dx} = \frac{2t - 3}{6t^2}$ so the numerator of this fraction gives:

$\frac{d}{dt}\left(\frac{dy}{dx}\right) = \frac{d}{dt}\left[\frac{2t - 3}{6t^2}\right]$ which is a quotient of two functions of t.

Step 5: Use the quotient rule of differentiation with $u = 2t - 3$ and $v = 6t^2$

$$\Rightarrow \frac{du}{dt} = 2 \text{ and } \frac{dv}{dt} = 12t$$

$$\text{so } \frac{d}{dt}\left(\frac{dy}{dx}\right) = \frac{2(6t^2) - 12t(2t - 3)}{36t^4}$$

Continued on the next page

Method notes

The numerator has not been simplified because at the stationary point we found earlier that $2t-3=0$

Method notes

To classify a stationary point you need to find the sign of $\frac{d^2y}{dx^2}$ at that point. This work was covered in Core 2

Therefore the stationary point is a local minimum.

Essential notes

$\frac{d^2y}{dx^2}$ means differentiate $\frac{dy}{dx}$ with respect to x. Using mathematical notation this means $\frac{d}{dx}\left(\frac{dy}{dx}\right)$ which can also be written as $\frac{d}{dt}\left[\left(\frac{dy}{dx}\right)\right]\left(\frac{dt}{dx}\right)$ using the chain rule $\frac{dt}{dx}=\frac{1}{\frac{dx}{dt}}$. The quotient and chain rules were covered in Core 3.

Essential notes

You do not need to know a proof of the area under a curve formula in terms of a parameter but you do need to know how to apply it.

Step 6: Substitute $\frac{dx}{dt}=6t^2$ from step 1 and the result from step 5 into the formula for $\frac{d^2y}{dx^2}$

$$\text{So } \frac{d^2y}{dx^2}=\frac{\frac{d}{dt}\left(\frac{dy}{dx}\right)}{\frac{dx}{dt}}=\frac{12t^2-12t(2t-3)}{216t^6}$$

At the stationary point $2t-3=0$ so $\frac{d^2y}{dx^2}=\frac{12t^2}{216t^6}>0$ at $t=\frac{3}{2}$

Finding areas under curves described parametrically

One of the applications of integration is to find the area A under a graph.

The formula is given by $A=\int y\mathrm{d}x$ and the upper and lower limits are 'x' values.

If the equation of the curve is given in terms of parametric equations with $y=\mathrm{g}(t)$ and $x=\mathrm{f}(t)$ then the chain rule leads to $A=\int y\frac{dx}{dt}\,dt$ and the upper and lower limits are 't' values.

Example

A curve has parametric equations $x=1+2t$ and $y=t^2+4$

Evaluate the integral $A=\int_1^3 y\,dx$

Answer

The curve is given parametrically so we must use the formula

$A=\int y\frac{dx}{dt}\,dt$ and the upper and lower limits must be t values not x values.

Step 1: The lower limit given is when $x=1$ so find the 't' lower limit

$1+2t=1\Rightarrow t=0$

Step 2: The upper limit given is when $x=3$ so find the 't' upper limit

$1+2t=3\Rightarrow t=1$

Step 3: Find $\frac{dx}{dt}$ so given $x=1+2t\Rightarrow\frac{dx}{dt}=2$

Step 4: Using the 't' limits from steps 1 and 2 and the result from step 3 rewrite and evaluate

$$\int_1^3 y\,\mathrm{d}x = \int_0^1 y\frac{\mathrm{d}x}{\mathrm{d}t}\,\mathrm{d}t = \int_0^1 (t^2+4)\times 2\mathrm{d}t$$

$$= \int_0^1 2t^2 + 8\mathrm{d}t = \left[\frac{2t^3}{3} + 8t\right]_0^1 = \frac{26}{3} \text{ square units.}$$

Example

A curve defined parametrically by the equations $x = t^2$, $y = t^3$ where $t \geq 0$ is a parameter is shown in the diagram.

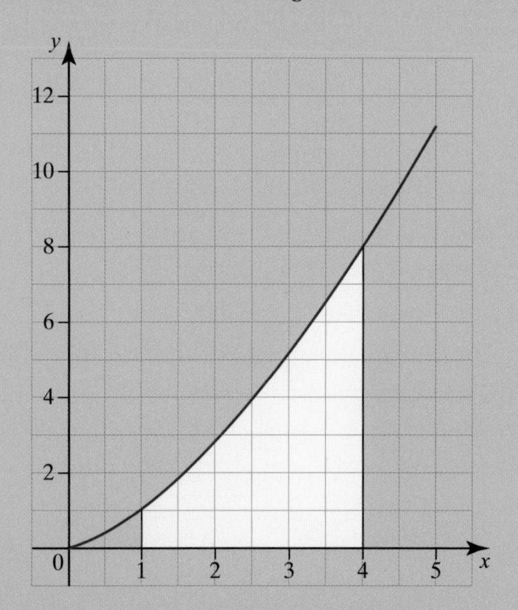

Fig. 3.4
Graph of the curve with parametric equations $x = t^2$, $y = t^3$

Find the area of the shaded region between the graph, the x-axis and the lines $x = 1$ and $x = 4$

Answer

The curve is given parametrically so we must use the formula

$A = \int y\frac{\mathrm{d}x}{\mathrm{d}t}\,\mathrm{d}t$ and the upper and lower limits must be t values not x values.

The area of the shaded region is given by the integral $\int_1^4 y\mathrm{d}x$

Step 1: The upper limit is when $x = 4$ so find the 't' upper limit:

$t^2 = 4 \Rightarrow t = 2$ or -2 but dismiss $t = -2$ as $t \geq 0$

Continued on the next page

Step 2: The lower limit is when $x = 1$ so find the 't' lower limit

$t^2 = 1 \Rightarrow t = 1$ or -1 but dismiss $t = -1$ as $t \geq 0$

Step 3: Find $\frac{dx}{dt}$ so given $x = t^2 \Rightarrow \frac{dx}{dt} = 2t$

Step 4: Using the 't' limits from steps 1 and 2 and the result from step 3 rewrite and evaluate

$$\int_1^4 y\,dx = \int_1^2 y\frac{dx}{dt}\,dt = \int_1^2 t^3 \times 2t\,dt = \int_1^2 2t^4\,dt$$

$$= \left[\frac{2t^5}{5}\right]_1^2 = \frac{2^6}{5} - \frac{2}{5} = \frac{62}{5}$$

Therefore the area of the shaded region is $\frac{62}{5} = 12.4$ square units.

Stop and think answers

1a) i) *Given the parametric equations* $x = 5t$ *and* $y = 2 - 3t^2$ *then*

differentiating with respect to the parameter $t \Rightarrow \frac{dx}{dt} = 5$ *and* $\frac{dy}{dt} = -6t$

Therefore using the chain rule $\frac{dy}{dx} = \frac{dy}{dt} \times \frac{dt}{dx}$ *and* $\frac{dt}{dx} = \frac{1}{\frac{dx}{dt}}$

$$\text{so } \frac{dy}{dx} = -6t \times \frac{1}{5} = \frac{-6t}{5}$$

which is the gradient of the tangent to the curve at any point.

The tangent and normal are perpendicular lines so their gradients multiply to give -1

Therefore the gradient of the normal at any point is $\frac{5}{6t}$

ii) *At the point* $(5, -1)$ *then if* $x = 5t$ *then* $5 = 5t$ *so* $t = 1$ *and if* $y = 2 - 3t^2$ *then* $-1 = 2 - 3t^2$ *so* $t^2 = 1$ *hence* $t = 1$ *(so that* $x = 5 \times 1 = 5$*)*

therefore from part (i) the gradient of the normal is $\frac{5}{6(1)} = \frac{5}{6}$

b) *To find* $\frac{d^2y}{dt^2}$ *means we must differentiate* $\frac{dy}{dt}$ *with respect to* t

and if $\frac{dy}{dt} = -6t$ *then* $\frac{d^2y}{dt^2} = -6$ *at any point on the curve.*

c) *To find* $\frac{d^2x}{dt^2}$ *means we must differentiate* $\frac{dx}{dt}$ *with respect to* t

and if $\frac{dx}{dt} = 5$ *then* $\frac{d^2x}{dt^2} = 0$ *at any point on the curve.*

Implicit differentiation

Different forms of equations of curves

Equations of a curve can be given in various forms as follows:

$y = x^2 + 3x - 5$ is an example of an equation where $y = f(x)$; this is called an **explicit equation** for a curve **because y is the subject of the equation, or y is given explicitly in terms of x.** In this case, y is the dependent variable which means we can calculate the value of y if we are given a value of x, the independent variable.

$x = t$ and $y = t^2 - 1$ is an example of a pair of **parametric equations** in which the curve $y = f(x)$ is defined in terms of a parameter t. To find the explicit formula for y in terms of x we must eliminate t between the two parametric equations.

$(x - 2)^2 + (y + 3)^2 = 16$ is an example of the equation of a circle. This is called an **implicit equation** for a curve because neither of the two variables, x and y is given as the subject of the equation.

Finding the gradient function given different forms of the equations of a curve

Example

Find the gradient function for the curve given explicitly as $y = x^2 + 3x - 5$

Answer

$y = x^2 + 3x - 5$ is an explicit equation and to find the gradient function $\frac{dy}{dx}$ we need to differentiate y with respect to x

so $\frac{dy}{dx} = 2x + 3$ which is the gradient function.

Method notes

Elimination of the parameter between the two parametric equations was covered in Chapter 3

An alternative method for finding the gradient function is to use the chain rule $\frac{dy}{dx} = \frac{dy}{dt} \times \frac{dt}{dx}$.

This was covered in Chapter 3

Example

Find the gradient function for the curve described parametrically by the equations $x = t$ and $y = t^2 - 1$

Answer

To find the gradient function we find the explicit equation for y in terms of x by eliminating t between the two parametric equations, then differentiate y with respect to x.

If $x = t$ then $y = x^2 - 1$ so $\frac{dy}{dx} = 2x$ which is the gradient function.

Example

Explain how to find the gradient function for the circle $(x-2)^2+(y+3)^2=16$

Answer

To find the gradient function we rearrange the implicit equation into an explicit equation for y in terms of x, then we can differentiate y with respect to x.

Given $(x-2)^2+(y+3)^2=16$ we can rewrite this as $(y+3)^2=16-(x-2)^2$

Taking the square root of both sides of the equation gives

$y+3=\pm\sqrt{16-(x-2)^2}$ so $y=-3\pm\sqrt{16-(x-2)^2}$

$y=-3\pm(16-(x-2)^2)^{\frac{1}{2}}$

This is now an explicit equation for the circle.

So $\frac{dy}{dx}$ is then found by using the chain rule (twice).

However, there is a much easier and quicker method called **implicit differentiation** as explained below!

Example

Find the gradient function $\frac{dz}{dx}$ given $z=y^n$ where $y=f(x)$ and n is a constant.

Answer

$z=y^n$ means z is a function of y and y is a function of x. Therefore z is a function of a function of x.

Step 1: Use the chain rule to differentiate z (which is a function of a function of x) with respect to x:

$$\frac{dz}{dx}=\frac{d(y^n)}{dx}=\frac{d(y^n)}{dy}\times\frac{dy}{dx}$$

Step 2: Differentiate z with respect to y so if $z=y^n$ then $\frac{dz}{dy}=ny^{n-1}$

Step 3: Combine the results from steps 1 and 2:

$$\frac{d(y^n)}{dx}=ny^{n-1}\frac{dy}{dx}$$

Method notes

The chain rule for differentiating a function of a function was covered in Core 3

The example above leads to a general result using $f'(y)$ notation:

the derivative of f(y) with respect to x is $\mathbf{f'(y)\frac{dy}{dx}}$

This is the method of implicit differentiation.

Essential notes

Implicit differentiation is so called because an implicit function f(y) has been differentiated without first finding the explicit relationship between the variables y and x.

You must learn the implicit differentiation method.

$f'(y)\frac{dy}{dx}$ means that we have differentiated f(y) with respect to y to give $f'(y)$ and then differentiated y with respect to x to give $\frac{dy}{dx}$.

Example

Differentiate x^2y with respect to x where $y = f(x)$.

Answer

Step 1: x^2y is a product of two functions of x so let 'u' $= x^2$ and 'v' $= y$ and differentiate with respect to x to give $\frac{du}{dx} = 2x$ and $\frac{dv}{dx} = \frac{dy}{dx}$

Step 2: Apply the product rule to give:

$$\frac{d(x^2y)}{dx} = v\frac{du}{dx} + u\frac{dv}{dx} = 2xy + x^2\frac{dy}{dx}$$

We now consider examples involving more general implicit equations.

Example

Find $\frac{dy}{dx}$ in terms of x and y for the implicit equation $x^4 + 2x + y^3 - 3xy = 10$

Answer

In this example the method is to differentiate each term in the expression with respect to x.

Step 1: Differentiate x^4 with respect to x to give $4x^3$

Step 2: Differentiate $2x$ with respect to x to give 2

Step 3: Differentiate y^3 with respect to x using the implicit method to give $3y^2\frac{dy}{dx}$

Step 4: Differentiate $3xy$ with respect to x using the product rule with 'u' $= 3x$ so $\frac{du}{dx} = 3$ and 'v' $= y$ so $\frac{dv}{dx} = \frac{dy}{dx}$ to give $3y + 3x\frac{dy}{dx}$

Step 5: Use the results from steps 1, 2, 3 and 4 to give

$$4x^3 + 2 + 3y^2\frac{dy}{dx} - \left(3y + 3x\frac{dy}{dx}\right) = 0$$

Step 6: Simplify to give $(4x^3 + 2 - 3y) + (3y^2 - 3x)\frac{dy}{dx} = 0$

so $(3y^2 - 3x)\frac{dy}{dx} = -(4x^3 + 2 - 3y)$

therefore $\frac{dy}{dx} = \frac{-(4x^3 + 2 - 3y)}{(3y^2 - 3x)} = -\frac{(4x^3 + 2 - 3y)}{3(y^2 - x)}$

Stop and think 1

Given the equation $(x-2)^2+(y+3)^2=16$ find $\frac{dy}{dx}$ in terms of x and y using the method of implicit differentiation.

Example

The equation of a curve is given by the implicit equation
$y^3-4x^2y+4x^2=16$

a) Find the slope of the tangent to the curve at the point (2, 4).

b) Find the equation of the tangent to the curve through the point (2, 4).

Answer

a) **Step 1:** Differentiate y^3 with respect to x to give $3y^2\frac{dy}{dx}$

Step 2: Differentiate $4x^2y$ with respect to x using the product rule with 'u' $=4x^2$ so $\frac{du}{dx}=8x$ and 'v' $=y$ so $\frac{dv}{dx}=\frac{dy}{dx}$ to give $8xy+4x^2\frac{dy}{dx}$

Step 3: Differentiate $4x^2$ with respect to x to give $8x$.

Step 4: Use the results from steps 1, 2 and 3 to give

$$3y^2\frac{dy}{dx}-\left(8xy+4x^2\frac{dy}{dx}\right)+8x=0$$

Step 5: Substitute $x=2$ and $y=4$ into the equation from step 4 to give

$$48\frac{dy}{dx}-\left(64+16\frac{dy}{dx}\right)+16=0$$

Step 6: Simplify the equation in step 5 to give

$$32\frac{dy}{dx}-48=0\Rightarrow\frac{dy}{dx}=1.5$$

Hence the slope of the tangent to the curve at the point (2, 4) is 1.5

b) The equation of the tangent to the curve at the point (2, 4) is the equation of a straight line with gradient 1.5 so $m=1.5$, $x_1=2$ and $y_1=4$

therefore $\frac{y-4}{x-2}=1.5\Rightarrow y=1.5x+1$

Hence the equation of the tangent is $y=1.5x+1$

Method notes

The slope of the tangent is give by the value of $\frac{dy}{dx}$ when $x=2$ and $y=4$

The method for finding $\frac{dy}{dx}$ is to differentiate

$y^3-4x^2y+4x^2=16$

term by term using the implicit method.

Method notes

To find the equation of the tangent at the point (2, 4) we use the general equation of a straight line passing through the point (x_1, y_1) and gradient m which is $\frac{y-y_1}{x-x_1}=m$

Exponential growth and decay

Consider the problem of the growth of a colony of bacteria in a laboratory experiment. A laboratory technician is told that the colony has a population of one million at midday on Sunday and during the next week the colony's population doubles every 24 hours. From this information she produces the following table of population values taken at midday each day.

Day	0	1	2	3	4	5	6	7
Population (in millions)	1	2	4	8	16	32	64	128

From this table of results we can see that if P is the population in millions and t is the day number (when the population is measured) then

after 1 day $P = 2^1$

after 2 days $P = 4 = 2^2$

after 3 days $P = 8 = 2^3$ so there is a relationship between the number of days and the population in millions.

The number of days is the 'power' or index of 2 so after t days $P = 2^t$

Therefore a formula which gives the population in millions after t days is $P = 2^t$.

This is an example of an **exponential growth model**. We say that 'as t increases P will increase exponentially'.

Cooling a hot liquid provides an example of another exponential model but this time it involves a reduction in the dependent variable (temperature) instead of the increase that we saw for population.

The following table shows the temperature of a flask of hot water measured in five minute intervals. The water was originally boiling with a temperature of 100°C.

Time (minutes)	0	5	10	15	20	25
Temperature (°C)	100	80	64	51.2	40.96	32.768

From this table of results we can see that if T is the temperature (measured in °C) and t is the time in minutes when the temperature is measured then

You start measuring when $t = 0$ then $T = 100$

when $t = 5$ then $T = 80 = 100 \times (0.8)^1 = 100 \times (0.8)^{\frac{5}{5}}$

when $t = 10$ then $T = 64 = 100 \times (0.8)^2 = 100 \times (0.8)^{\frac{10}{5}}$

so there is a relationship between the starting temperature, (0.8) and the time in minutes.

The power of (0.8) is the number of minutes at which the temperature is measured, divided by 5

Therefore a formula for the temperature (in °C) in terms of time t (in minutes) is $T = 100 \times 0.8^{\frac{t}{5}}$

This is an example of an **exponential decay model**.

Summary

The function $y = a^x$ describes exponential growth or decay where y and x are variables and a is a constant.

If $a > 1$ then the variable y will increase (exponential growth).

If $a < 1$ then the variable y will decrease (exponential decay).

If $a = 1$ then y is a constant.

The derivative $\frac{dy}{dx}$ of the function $y = a^x$ gives a measure of the rate of change of growth or decay.

Essential notes

Exponential functions were covered in Core 3

An exponential function was defined as $y = a^x$ where y and x are variables. Any letters can be used to represent the variables. In the exponential growth model we use P and t as the variables. In the exponential decay model we use T and t as the variables.

Derivative of the exponential function $y = a^x$

Given the exponential function $y = a^x$ where x and y are variables and a is a constant, if we take logs of both sides this gives $\ln y = x \ln a$ which is an implicit formula involving x and y as y is a function of x.

To find $\frac{dy}{dx}$ which is the derivative of y we use the method of implicit differentiation.

Step 1: Differentiate $\ln y = x \ln a$ implicitly with respect to x

$$\frac{1}{y}\frac{dy}{dx} = \ln a$$

Step 2: Multiply both sides of the equation in step 1 by y

$$\frac{dy}{dx} = y \ln a$$

Step 3: Substitute $y = a^x$ into the equation in step 2

$$\frac{dy}{dx} = a^x \ln a$$

General result: if $y = a^x$ then $\frac{dy}{dx} = a^x \ln a$

Essential notes

'Use the method of implicit differentiation' is often abbreviated to 'differentiate implicitly'.

Method notes

The derivative of $\ln y$ with respect to y is $\frac{1}{y}$.

a is a constant so $\ln a$ is also a constant.

Exam tips

You must remember this general result. In the special case where the base $a = e$ then $y = e^x$ so $\frac{dy}{dx} = e^x \ln e = e^x$ because $\ln e = 1$

Example

Find $\frac{dy}{dx}$ if $y = 3^x$.

Answer

Step 1: Compare $y = 3^x$ with the function $y = a^x$ so $a = 3$

Step 2: Use the general result for the differentiation of $y = a^x$ with $a = 3$

so $\frac{dy}{dx} = 3^x \ln 3$

Rates of change

You are familiar with finding the gradient function $\frac{dy}{dx}$ for different curves. The gradient function is an example of a rate of change of the two variables y and x. The symbol $\frac{dy}{dx}$ means the rate at which y changes as x changes (and it gives the gradient of the tangent to the curve at any point on the curve). In examples where we have used the two letters v and t to represent velocity and time respectively then by differentiation we would obtain $\frac{dv}{dt}$.

The symbol $\frac{dv}{dt}$ means the rate at which v changes as t changes (and it gives the rate at which the velocity changes as time changes or the acceleration at a particular time).

You can use any two letters to represent different variables and in examples where we have the exponential function $y = a^x$, with y and x as the variables and a is a constant then the symbol $\frac{dy}{dx}$ means the rate at which y changes as x changes (and it gives the rate of change of growth or decay depending on the value of a).

The use of rates of change in problem solving is explained in the following examples.

Example

The temperature T of a flask of hot water cooling from 100°C at time t minutes is given by the equation $T = 100(0.8)^{0.2t}$.

Find the rate of change of the temperature after 15 minutes.

Answer

Given $T = 100(0.8)^{0.2t}$ the two variables are T and t.

To find the rate of change of temperature (T) after the given time (t) means we must find $\frac{dT}{dt}$ when $t = 15$

Step 1: Compare with the general exponential equation $y = a^x$

so $x = 0.2t$, $a = 0.8$, $y = 0.8^x$ and $T(x) = 100y$.

Step 2: Differentiate $x = 0.2t$ so $\frac{dx}{dt} = 0.2$

Step 3: Use general differentiation result for '$y = a^x$ so $\frac{dy}{dx} = a^x \ln a$' with

$y = 0.8^x$ so $\frac{dy}{dx} = 0.8^x \times \ln 0.8$

$T(x) = 100y \Rightarrow \frac{dT}{dx} = 100 \times \frac{dy}{dx} = 100 \times 0.8^x \times \ln 0.8$

Step 4: Use the chain rule and the results from steps 2 and 3 to find $\frac{dT}{dt}$

with $x = 0.2t$ so $\frac{dT}{dt} = \frac{dT}{dx} \times \frac{dx}{dt}$ gives

$\frac{dT}{dt} = 100 \times (0.8)^{0.2t} \times (\ln 0.8) \times 0.2 = -4.46(0.8)^{0.2t}$ °C per min

Step 5: After 15 minutes means when $t = 15$ so substitute $t = 15$ into the equation from step 4 therefore $\frac{dT}{dt} = -4.46 \times (0.8)^3 = -2.28$ °C per min.

Hence the hot water is cooling (signified by the – sign) at a rate of 2.28 °C per min.

Differential equations

Applied mathematicians, scientists or engineers often have to translate a physical situation into a set of equations which will describe that situation. It is usually impossible to describe the physical situation fully in terms of mathematical equations and it is often necessary to simplify the situation so that progress can be made. Once a set of equations is formed it is possible to compare the outcome from the mathematical equations with data collected from the real system.

The set of equations which describes a physical situation is called a **mathematical model** of the situation.

Common mathematical models which describe the laws of nature are called **differential equations**.

Essential notes

Direct proportion was covered in Core 1

If y varies directly as x we write it as $y \propto x$ and this means $y = kx$ where k is a constant to be determined.

For example, in physics, $\frac{dN}{dt} = -kN$ describes the decay of radioactive particles where N is the number of particles at a particular time t and k is a positive constant. The equation tells us that the rate at which N is changing as the time changes is directly proportional to the number (N) of particles remaining, at any given time (t).The symbol k is the constant of proportionality and the negative sign tells us that the number of particles is decreasing.

Definitions

A differential equation is an equation expressed in terms of an independent variable, a dependent variable and (some of) its derivatives. The differential equation is classified according to the 'order' of the highest derivative in that equation.

$\frac{dN}{dt} = -kN$ is classified as (or called) a **first order differential equation** because it involves a only first derivative.

$\frac{d^2x}{dt^2} + \frac{dx}{dt} = a^2 - x^2$ is called a **second order differential equation** because the 'highest' derivative is a second derivative $\frac{d^2x}{dt^2}$.

The work covered in this Chapter is about first order differential equations formed from physical situations which have been described in words.

Methods of solution of these equations are explored later in this chapter and in Chapter 5

Example

A wet and porous sheet of cloth gains moisture in a damp environment at a rate proportional to the moisture content which is y and is dependent on time t.

Write down an equation for the rate of change of moisture content of the cloth.

Answer

The cloth becomes more moist as the time changes so the two variables are the moisture content y and the time t.

The rate of change of these two variables is the rate of change of moisture content as the time changes and the symbol is $\frac{dy}{dt}$.

If this rate of change increases at a rate proportional to y this means $\frac{dy}{dt} \propto y$

so $\frac{dy}{dt} = ky$ where k is a positive constant. k must be positive as it is an increasing rate.

Example

The population of a given country is increasing due to births and deaths at a constant rate of $k\%$ of the current population per year.

The population is also decreasing at a constant rate of N people per year by emigration.

Write down an equation for the rate of growth of population.

Answer

Let the population of the country be P and let the time be t years. These are the two variables and the symbol for the rate of change of population as time changes is $\frac{dP}{dt}$.

Step 1: The rate of change of the population $\frac{dP}{dt}$ increases because of births and deaths and decreases because of emigration, so

$$\frac{dP}{dt} = \text{rate of increase} - \text{rate of decrease}$$

Step 2: Work out the constant rate of increase which was given as $k\%$ of the current population (P) so rate of increase $= \frac{k}{100}P$

Step 3: Work out the rate of decrease which was given as the constant rate of N people per year so rate of decrease $= N$

Step 4: Substitute the results of steps 2 and 3 in the equation of step 1

so $\frac{dP}{dt} = \frac{k}{100}P - N$

This is a first order differential equation for the rate of growth of the population.

Example

The slope of the tangent to a curve at the point (x, y) is given by $(x + y)$.

Formulate a first order differential equation for finding the equation of the graph.

Answer

The slope of the tangent to any curve $y = f(x)$ is the gradient function of the graph which is $\frac{dy}{dx}$. We are given that the slope is $(x + y)$ so we write $\frac{dy}{dx} = (x + y)$ which is a first order differential equation.

Example

The area, A, of a circular oil slick is increasing at a rate proportional to its radius, r.

a) Write down an expression for $\frac{dA}{dt}$.

b) Find an expression for $\frac{dA}{dr}$ in terms of r.

c) Hence formulate a differential equation involving $\frac{dr}{dt}$ and interpret the result.

Answer

a) The rate of change of area, A, with respect to time, t, $\frac{dA}{dt}$ is proportional to its radius, r so $\frac{dA}{dt} \propto r$ therefore $\frac{dA}{dt} = kr$ where k is a positive constant since the area is increasing.

b) A circular oil slick has area $A = \pi r^2$ so $\frac{dA}{dr} = 2\pi r$

c) Using the chain rule, $\frac{dr}{dt} = \frac{dr}{dA} \times \frac{dA}{dt} = \frac{\frac{dA}{dt}}{\frac{dA}{dr}} = \frac{kr}{2\pi r} = \frac{k}{2\pi}$

Since k is a constant then the time of change of the radius is constant.

Example

A cylindrical tank contains 100 litres of brine in which 10 kg of salt is dissolved in water.

A solution of brine containing 2 kg of salt per litre flows into the tank at 5 litres per minute.

The mixture is thoroughly stirred and drawn off at 4 litres per minute.

Find an expression for the amount of salt in the tank at any time.

Answer

Let $q(t)$ be the amount of salt in the tank at time t.

The rate at which the salt is accumulating in the tank is given by:

$\frac{dq}{dt}$ = amount of salt entering per minute − amount of salt leaving per minute

A solution of brine containing 2 kg of salt per litre flows into the tank at 5 litres per minute so the amount of salt entering is 10 kg per minute.

Amount of salt entering per minute = 10 kg per minute

5 litres of fluid is added per minute and 4 litres of brine is drawn off per minute so after t minutes the tank has increased its volume by $(5-4)t = t$ litres of brine.

The tank contained 100 litres of brine initially so after t minutes the volume of liquid in the tank is $(100+t)$ litres.

So the concentration of salt in the tank after t minutes is $\frac{q}{(100+t)}$ kg per litre and the amount of salt leaving $= \frac{4q}{(100+t)}$ kg per minute

so $\frac{dq}{dt} = 10 - \frac{4q}{(100+t)}$ which is a first order differential equation.

Solving differential equations by direct integration

You have already solved differential equations of the type $\frac{dy}{dx} = f(x)$ in the chapters on integration in previous modules. The method of solution of $\frac{dy}{dx} = f(x)$ is by simple integration. Other methods are explored in Chapter 5

Example

The slope of the tangent to a curve at the point (x, y) is given by $(x^2 + x - 1)$. The curve passes through the point $(0, 1)$.

a) Formulate a differential equation for finding the equation of the curve.

b) Solve the differential equation to find the equation of the curve.

Answer

a) The slope of the tangent or gradient function of a graph is $\frac{dy}{dx}$

so we write $\frac{dy}{dx} = (x^2 + x - 1)$ which is a first order differential equation.

Method notes

This is called the **general solution of the differential equation**. It represents a family of curves, one for each value of the constant c.

The graph passes through the point $(0, 1)$ so $y = 1$ when $x = 0$ which we call an initial (or boundary) condition.

Continued on the next page

b) $\frac{dy}{dx} = (x^2 + x - 1)$

By direct integration $y = \int (x^2 + x - 1)dx = \frac{x^3}{3} + \frac{x^2}{2} - x + c$

But we have the initial condition $x = 0, y = 1$

so $1 = \frac{0^3}{3} + \frac{0^2}{2} - 0 + c \Rightarrow c = 1$

Therefore the equation of the curve passing through the point (0, 1) is

$y = \frac{x^3}{3} + \frac{x^2}{2} - x + 1$

Method notes

This is called the **particular solution of the differential equation**. It is the one curve that has the given slope and passes through the given point.

Example

The area, A, of a circular oil slick is increasing at a rate proportional to its radius, r (in meters).

A differential equation for the rate of increase of the radius is $\frac{dr}{dt} = \frac{k}{2\pi}$ where k is a positive constant and t is the time in hours.

Initially when the oil slick was formed it had a radius of 5 metres. After 2 hours the radius was 12 metres.

Find the value of k and an expression for the radius, r, as a function of t.

Answer

Given the first order differential equation $\frac{dr}{dt} = \frac{k}{2\pi}$ where k is a constant (and also 2π) we can solve it by direct integration.

Step 1: Integrate $\frac{dr}{dt} = \frac{k}{2\pi}$ with respect to t so $r = \frac{k}{2\pi}t + c$ where c is the constant of integration.

Given that the initial radius is 5 m this means that when $t = 0$ $r = 5$ which is an initial (or boundary) condition.

Step 2: Substitute $t = 0$ and $r = 5$ into the equation from step 1 so

$5 = \frac{k}{2\pi}(0) + c$ and therefore $c = 5$

Step 3: Substitute the result from step 2 into the equation in step 1 so

$$r = \frac{k}{2\pi}t + 5$$

Given that after 2 hours the radius was 12 metres means that when $t = 2$, $r = 12$ (which is a boundary condition).

Step 4: Substitute $t = 2$ and $r = 12$ into the equation in step 3 so

$$12 = \frac{k}{2\pi}(2) + 5 \Rightarrow k = 7\pi$$

Step 5: Substitute the result from step 4 into the equation in step 3 so

$r = \frac{7}{2}t + 5$ which is an expression for the radius, r, as a function of t.

Stop and think answers

Given $(x-2)^2 + (y+3)^2 = 16$ then we differentiate implicity with respect to x:

$(x-2)^2$ is a function of a function of x so using the chain rule to differentiate with respect to x this differentiates to $2(x-2)^1 \times 1$ which is $2(x-2)$

$(y+3)^2$ is a function of a function of y and y is a function of x so using the chain rule to differentiate with respect to y gives $2(y+3)^1 \times 1$ then we need to differentiate y with respect to x which is $\frac{dy}{dx}$ therefore the full result is $2(y+3)\frac{dy}{dx}$

16 is a constant and differentiates to 0 so combining these results gives

$$2(x-2) + 2(y+3)\frac{dy}{dx} = 0$$

Rearranging gives $\frac{dy}{dx} = \frac{-2(x-2)}{2(y+3)} = \frac{2-x}{y+3}$

Essential notes

If the function f(x) is negative between $x = a$ and $x = b$ then the value of the integral is negative. So in this case the area between the curve $y = f(x)$, the x-axis and the lines $x = a$ and $x = b$ is $\int_a^b f(x)dx$ i.e. the numerical value of the integral.

You have met the topic of integration in Core 1 and Core 2 and the important results that you should know from this previous work are as follows:

- the rule for **indefinite** integral of a power of x is
  $\int x^n dx = \frac{1}{n+1}x^{n+1} + c, n \neq -1$ where c is a constant of integration;
- for a function f(x) which is positive in the interval $a < x < b$ the the area between the curve $y = f(x)$, the x-axis and the lines $x = a$ and $x = b$ is
  $$\int_a^b f(x)dx$$
- integration is the reverse process of differentiation so that for powers of x
  if $\frac{dy}{dx} = x^n$ then $y = \frac{1}{n+1}x^{n+1} + c, n \neq -1$ so $\int x^n dx = \frac{1}{n+1}x^{n+1} + c$
  which can also be written as $\int \frac{dy}{dx} dx = y$
- more generally
  if $y = \int f(x)dx = g(x)$ then $\frac{dg(x)}{dx} = f(x)$

Essential notes

Remember the result $\int \frac{dy}{dx} dx = y$ as it will be referred to later in this Chapter.

In words this means if the integral of a function of x (which we will call f(x)) is g(x) + c then the derivative of g(x) + c with respect to x is f(x).

Essential notes

The function to be integrated, f(x), is called the **integrand**.

Integrating standard functions

The link between integration and differentiation provides a method for integrating certain standard functions such as exponential functions, trigonometric functions and the reciprocal function $\frac{1}{x}$.

Integrating exponential functions

Let $\int e^{kx}dx = g(x)$ where k is a constant.

If we are to work out $\int e^{kx}dx$ it means we are looking for a function g(x) which, when differentiated, will give us a function of x, that is $f(x) = e^{kx}$.

Using mathematical notation this means $\frac{dg(x)}{dx} = f(x) = e^{kx}$.

You know that $\frac{d(e^{kx})}{dx} = ke^{kx}$.

Dividing by k gives $\frac{d\left(\frac{e^{kx}}{k}\right)}{dx} = e^{kx}$

Comparing $\frac{dg(x)}{dx} = f(x) = e^{kx}$ and $\frac{d\left(\frac{e^{kx}}{k}\right)}{dx} = e^{kx}$ we deduce that

$g(x) = \frac{e^{kx}}{k}$ so

$\int e^{kx}dx = \frac{e^{kx}}{k} + c$ where c is the constant of integration.

Exam tips

You need to remember this formula as it is not in the formula book.

Example

Find the following integrals a) $\int e^{2x}dx$ b) $\int_0^2 e^{3x}dx$

Answer

a) **Step 1:** Compare $\int e^{2x}\,dx$ with the standard result $\int e^{kx}dx = \frac{e^{kx}}{k} + c$

so $k = 2$

Step 2: Substitute $k = 2$ into the equation in step 1:

$$\int e^{2x}dx = \frac{e^{2x}}{2} + c$$

b) **Step 1:** Compare $\int_0^2 e^{3x}dx$ with the standard result $\int e^{kx}dx = \frac{e^{kx}}{k} + c$

so $k = 3$

Here we must ignore c, the constant of integration, as $\int_0^2 e^{3x}dx$ is a definite integral.

Step 2: Substitute $k = 3$ into the equation from step 1, ignoring c and stating the upper and lower limits:

$$\int_0^2 e^{3x}dx = \left[\frac{e^{3x}}{3}\right]_0^2$$

Step 3: Evaluate the integral in step 2 using the upper and lower limits:

$$\int_0^2 e^{3x}dx = \left[\frac{e^{3x}}{3}\right]_0^2 = \frac{e^6}{3} - \frac{e^0}{3} = \frac{e^6 - 1}{3}$$

Therefore $\int_0^2 e^{3x}dx = \frac{e^6 - 1}{3}$

Essential notes

Definite integrals were explained in Chapter 7 of Core 2

Integrating trigonometric functions

Consider the integral $\int \sin x\,dx = g(x)$

You know that the derivative of $\cos x = -\sin x$ so using mathematical notation this means $\frac{d(\cos x)}{dx} = -\sin x$ and mutiplying both sides of this equation by -1 gives $\frac{d(-\cos x)}{dx} = \sin x$.

Using the link between integration and differentiation as explained in the last example this means that $\int \sin x\,dx = -\cos x + c$ where c is a constant of integration.

A similar approach using the results of trigonometric differentiation, which was covered in Chapter 4 of Core 3, gives the following standard results for integration where c is the constant of integration.

Essential notes

You must learn the link between integration and differentiation which is: if the integral of a function is $g(x)$ then it means we are looking for a function $g(x)$ which, when differentiated, will give us the function of x which was to be integrated. This function of x is called the integrand.

Standard results for integration of trigonometric functions

$$\int \sin x\,dx = -\cos x + c$$

$$\int \cos x\,dx = \sin x + c$$

$$\int \sec^2 x\,dx = \tan x + c$$

$$\int \operatorname{cosec} x \cot x\,dx = -\operatorname{cosec} x + c$$

$$\int \sec x \tan x\,dx = \sec x + c$$

$$\int \operatorname{cosec}^2 x\,dx = -\cot x + c$$

Exam tips

You need to remember the standard integration results for $\sin x$ and $\cos x$ as they are not in the formula book. In particular be careful to include the –ve sign in the answer for integral of $\sin x$.

Example

Find the following integrals

a) $\int (2\sin x - 3\cos x)\,dx$

b) $\int_0^{\frac{\pi}{4}} \left(\frac{1}{\cos^2 x} + \frac{\sin x}{\cos^2 x}\right) dx$

Answer

a) Use the standard results above and integrate each term separately so $\int 2\sin x - 3\cos x\,dx = \int 2\sin x\,dx - \int 3\cos x\,dx = -2\cos x - 3\sin x + c$ where c is the constant of integration.

b) **Step 1:** Rewrite each term to be integrated so that they can be recognised as standard integrals so $\int_0^{\frac{\pi}{4}} \frac{1}{\cos^2 x} + \frac{\sin x}{\cos^2 x}\,dx$

Method notes

Any constant multiplier of a function of x is unaffected by the integration process. In $\int 2\sin x\,dx$, 2 is the constant multiplier.

This was covered in Core 1

$$= \int \frac{1}{(\cos x)^2}\,dx + \int \left(\frac{\sin x}{\cos x}\right)\left(\frac{1}{\cos x}\right)dx$$

$$= \int \sec^2 x\,dx + \int \tan x \sec x\,dx$$ ignoring the c (definite integral).

Step 2: Use the standard results of integration for the equation in step 1, state and use the upper and lower limits:

$$\int \sec^2 x\,dx + \int \tan x \sec x\,dx = [\tan x + \sec x]_0^{\frac{\pi}{4}}$$

$$= \left[\tan\frac{\pi}{4} + \sec\frac{\pi}{4}\right] - [\tan 0 + \sec 0]$$

$$= \left[1 + \sqrt{2}\right] - [0 + 1] = \sqrt{2}$$

Integrating the reciprocal function $\frac{1}{x}$

Consider the integral $\int \frac{1}{x}dx = g(x)$. Using the link between integration and differentiation we are looking for a function $g(x)$ with the property $\frac{dg(x)}{dx} = f(x) = \frac{1}{x}$.

From the work covered in Chapter 4 of Core 3 you know that

$$\frac{d(\ln x)}{dx} = \frac{1}{x} \text{ so } \int \frac{1}{x}dx = \ln x + c$$

Essential notes

It is more usual to write $\int \frac{1}{x}dx = \ln|x| + c$. Using the modulus sign avoids difficulties with negative values of x.

Standard result for $\int \frac{1}{x}dx$

$$\int \frac{1}{x}dx = \ln|x| + c, x \neq 0$$

Exam tips

You need to remember this standard result as it is not in the formula book.

Example

Integrate the function $\frac{1}{x} + \frac{1}{x^2}$ with respect to x.

Answer

Step 1: Rewrite each term separately ready for using the standard results of integration:

$$\int \left(\frac{1}{x} + \frac{1}{x^2}\right)dx = \int \frac{1}{x}dx + \int \frac{1}{x^2}dx = \int \frac{1}{x}dx + \int x^{-2}dx$$

Step 2: Use the standard results for the equation in step 1

$$\text{so } \int \left(\frac{1}{x} + \frac{1}{x^2}\right)dx = \ln|x| + \frac{x^{-1}}{-1} = \ln|x| - \frac{1}{x} + c$$

Method notes

Be careful when integrating expressions of the form $\frac{1}{x^n}$.

It is only when $n = 1$ that you get a logarithm in the answer. For other values of n, use the standard rule of integration: $\int x^n dx = \frac{x^{n+1}}{n+1} + c$

Integrating linear transformations of functions

Essential notes

Linear transformations of trigonometric functions were covered in Core 3

You are often asked to integrate a function of the form $f(ax + b)$ where f is one of the basic functions above, and a and b are constants.

For example the integral $\int \sin(3x + 4)dx$ involves the integrand $\sin(3x + 4)$ which is a linear transformation of the function $\sin x$.

To evaluate integrals of this type we use an application of the chain rule as explained in the following example.

Example

Find the following integrals:

a) $\int \sin(3x + 4)dx$ b) $\int e^{2x-5}dx$ c) $\int \frac{1}{4x + 5}dx$

Answer

a) **Step 1:** Using the standard function in the integrand $\sin(3x + 4)$ as '$\sin x$', you know that $\int \sin x dx = -\cos x +$ a constant of integration.

Step 2: Let $g(x) = -\cos(3x + 4)$ and find a function $g(x)$ with the property that $\frac{dg(x)}{dx} = \sin(3x + 4)$. The function $\sin(3x + 4)$ is a composite function of x (or a function of a function of x) as explained in Core 3 so to differentiate use the chain rule:

therefore $\frac{dg(x)}{dx} = \sin(3x + 4) \times 3$

Step 3: Use the link between integration and differentiation in the equation in step 2 so $\int 3 \sin(3x + 4)dx = -\cos(3x + 4) + c$ where c is a constant of integration.

Step 4: Divide the equation in step 3 by 3 so

$$\int \sin(3x + 4)dx = -\frac{1}{3}\cos(3x + 4) + \frac{c}{3} = -\frac{1}{3}\cos(3x + 4) + k$$

where $k = \frac{c}{3}$ and is a constant.

so $\int \sin(3x + 4)dx = -\frac{1}{3}\cos(3x + 4) + k$

b) **Step 1:** Using the 'standard' function in the integrand e^{2x-5} as 'e^x' you know that $\int e^x dx = e^x +$ a constant of integration.

Step 2: Let $g(x) = e^{2x-5}$ and find a function $g(x)$ with the property that $\frac{dg(x)}{dx} = e^{2x-5}$. The function e^{2x-5} is a composite function of x (or a function of a function of x) so to differentiate use the chain rule:

$$\frac{dg(x)}{dx} = e^{2x-5} \times 2$$

Step 3: Use the link between integration and differentiation in the equation in step 2 so $2\int e^{2x-5}dx = e^{2x-5} + c$ where c is a constant of integration.

Step 4: Divide the equation in step 3 by 2 so

$\int e^{2x-5}dx = \frac{1}{2}e^{2x-5} + \frac{c}{2} = \frac{1}{2}e^{2x-5} + k$ where $k = \frac{c}{2}$ and is a constant

so $\int e^{2x-5}dx = \frac{1}{2}e^{2x-5} + k$

c) **Step 1:** Using the 'standard' function in the integrand $\int \frac{1}{x}dx$ as $\frac{1}{x}$ you know that $\int \frac{1}{x}dx = \ln|x| +$ a constant of integration

Step 2: Let $g(x) = \ln|4x + 5|$ and find a function $g(x)$ with the property that $\frac{dg(x)}{dx} = \frac{1}{4x + 5}$. The function $\ln(4x + 5)$ is a composite function of x (or function of a function of x) so to differentiate use the chain rule:

$$\frac{dg(x)}{dx} = \frac{1}{4x + 5} \times 4$$

Step 3: Use the link between integration and differentiation in the equation in step 2:

$4\int \frac{1}{4x + 5}dx = \ln|4x + 5| + c$ where c is a constant of integration.

Step 4: Divide the equation in step 3 by 4:

$\int \frac{1}{4x + 5}dx = \frac{1}{4}\ln|4x + 5| + \frac{c}{4} = \frac{1}{4}\ln|4x + 5| + k$ where $k = \frac{c}{4}$ and is a constant

so $\int \frac{1}{4x + 5}dx = \frac{1}{4}\ln|4x + 5| + k$

From the results in the worked example above you can probably identify a simple rule:

$\int f(ax + b)dx = \frac{1}{a}g(ax + b) + c$ where $\frac{dg}{dx} = f(ax + b)$ and a, b and c are constants.

Using trigonometric identities to simplify integrals

In Core 3 you met the following trigonometric identities which provide useful methods for transforming integrals into standard forms:

$\sin^2 x + \cos^2 x \equiv 1$

$\tan^2 x + 1 \equiv \sec^2 x$

$1 + \cot^2 x \equiv \text{cosec}^2 x$

$\sin 2x \equiv 2\sin x \cos x$

$\cos 2x \equiv 2\cos^2 x - 1 \equiv 1 - 2\sin^2 x \equiv \cos^2 x - \sin^2 x$

Identities are statements which are true for all values of the variable, which in the statements above is x.

Example

Find the following integrals

a. $\int \tan^2 x \mathrm{d}x$

b. $\int \sin 5x \cos 5x \mathrm{d}x$

c. $\int \cos^2 x \mathrm{d}x$

Answer

a) **Step 1:** Simplify the integrand $\tan^2 x$ into a standard form using a trigonometric identity so $\tan^2 x + 1 \equiv \sec^2 x$ gives $\tan^2 x = \sec^2 x - 1$

Step 2: Substitute the identity from step 1 into $\int \tan^2 x \,\mathrm{d}x$ so

$$\int \tan^2 x \mathrm{d}x = \int (\sec^2 x - 1)\mathrm{d}x = \int \sec^2 x \mathrm{d}x - \int 1 \mathrm{d}x$$

Step 3: Use the standard result for the integration of $\sec^2 x$ and the rule of integration for the constant 1

so $\int \tan^2 x \mathrm{d}x = \tan x - x + c$ where c is the constant of integration.

b) **Step 1:** Simplify the integrand $\sin 5x \cos 5x$ using the 'double angle formula':

$$\sin 2A = 2 \sin A \cos A \text{ with } A = 5x$$

$\Rightarrow \quad 2 \sin 5x \cos 5x = \sin 2(5x)$

therefore $\sin 5x \cos 5x = \frac{1}{2} \sin 10x$

Step 2: Use the standard result for the integration of $\sin kx$ where k is a constant which is $\int \sin kx \,\mathrm{d}x = -\frac{1}{k} \cos kx + c$ and integrate the equation from step 1 so.

$$\int \sin 5x \cos 5x \mathrm{d}x = \int \frac{1}{2} \sin 10x \mathrm{d}x = \frac{1}{2} \times (-\cos 10x) \times \frac{1}{10} + c$$

$= -\frac{1}{20} \cos 10x + c$ where c is the constant of integration.

Essential notes

The double angle formula $\sin 2A = 2 \sin A \cos A$ was covered in Core 3

Any letter or multiple of a letter can be used for the angle so if $A = 5x$ then

$\sin 2(5x) = 2 \sin 5x \cos 5x$

giving $\sin 10x = 2\sin 5x \cos 5x$

c) **Step 1:** Simplify the integrand $\cos^2 x$ using the 'double angle formula'

$\cos 2x = 2\cos^2 x - 1$ which gives $1 + \cos 2x = 2\cos^2 x$ so

$\dfrac{1 + \cos 2x}{2} = \cos^2 x$ therefore $\dfrac{1}{2} + \dfrac{\cos 2x}{2} = \cos^2 x$

Step 2: Use the standard result for the integration of $\cos 2x$ which is

$\int \cos 2x \, dx = \dfrac{1}{2} \sin 2x$ and integrate the equation from step 1

so $\int \cos^2 x dx = \int \dfrac{1 + \cos 2x}{2} dx$

$= \int \dfrac{1}{2} + \dfrac{1}{2}\cos 2x dx = \dfrac{1}{2}x + \dfrac{1}{2}\left(\dfrac{1}{2}\sin 2x\right) + c$

so $\int \cos^2 x dx = \dfrac{1}{2}x + \dfrac{1}{4}\sin 2x = c$ where c is a constant of integration.

Stop and think 1

Find $\int \sin^2 x \, dx$.

Volumes of revolution

Finding the volume of a solid of revolution formed by rotating a curve about the *x*-axis

Figure 5.1 shows the area between the curve $y = f(x)$, the x-axis and the lines $x = a$ and $x = b$.

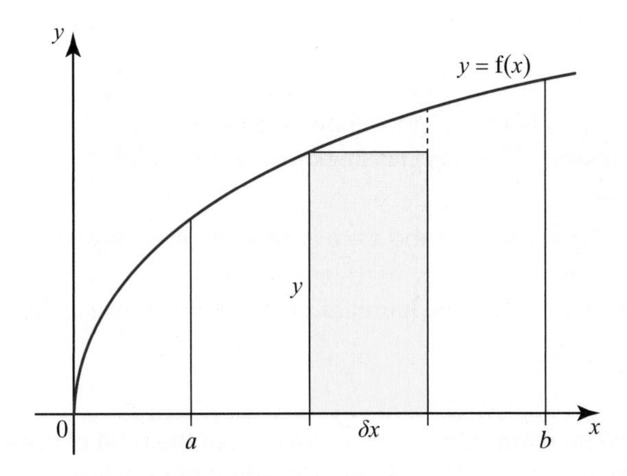

Fig. 5.1
A region divided into rectangular strips.

Essential notes

This idea arose in Core 2 where a region was divided into trapezia. This led to the **trapezium rule** for approximating an integral. Using integration as a summation also arose in Core 2

The area can be divided into rectangular strips with height y and width δx.

The value of the height of the rectangle depends on where it is drawn.

Essential notes

δx means a very small increase in x.

The area of the rectangular is $y\delta x$. If we fill the region with rectangles and add together (or sum) all the areas then the integral $\int_a^b y\,dx = \int_a^b f(x)\,dx$ can be found as the limit of the sum of these strips as $\delta x \to 0$ and the number of strips tends to infinity.

Suppose that we now rotate the rectangular strip of height y and width δx about the x-axis through 360°. The region swept out by the strip is a cylindrical disc of radius y and thickness δx as shown below in Figure 5.2.

Fig. 5.2
A disc formed by rotating the rectangle in Fig. 5.1

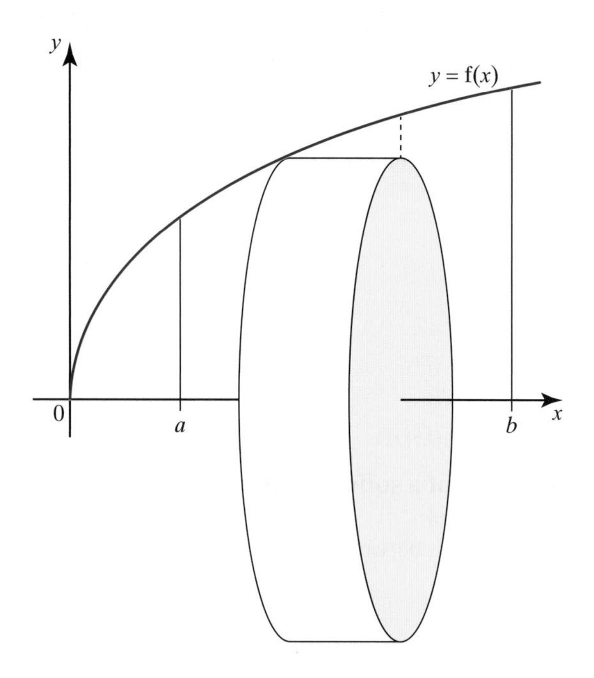

The cross section of the cylindrical disc is a circle so the disc has cross sectional area πy^2. This disc has thickness δx, therefore the volume of the disc is $\pi y^2 \delta x$.

If the region between $x = a$ and $x = b$ is divided into many such discs then the integral $\int_a^b \pi y^2\,dx$ can be found as the limit of the sum of these discs as

$\delta x \to 0$

This integral therefore represents the **volume of the solid of revolution** found by rotating the curve through 360° about the x-axis over the interval $a \le x \le b$.

Essential notes

The volume of the disc is the area of cross section × thickness which is the area of a circle radius $y \times \delta x$ hence $\pi y^2\,\delta x$. The process of using integration as a summation was covered in Core 2

The formula is therefore: volume of revolution $= \int_a^b \pi y^2 dx$

This solid of revolution is shown below in Figure 5.3.

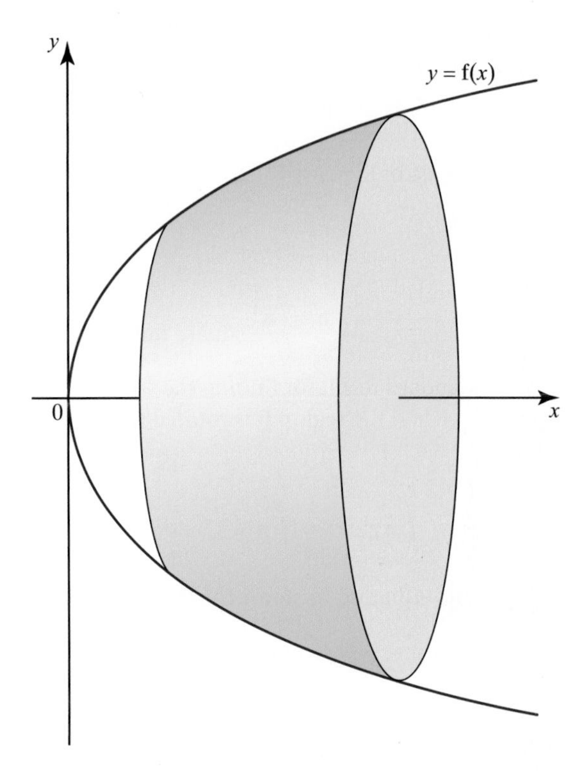

Exam tips

This result will be in the formula booklet which is provided in the examination. You will not be asked to prove the result but you must learn how to apply the formula.

Fig. 5.3
The volume of revolution formed by rotating the curve $y = f(x)$ through 360° about the x-axis.

Example

The region R is bounded by the curve $y = e^{2x+3}$, the x-axis and the lines $x = 0$ and $x = 1$

a. Find the area of the region R.

b. Find the volume of the solid of revolution formed when R is rotated through 360° about the x-axis.

Answer

c) **Step 1:** Use the standard result for finding the area of a region with the upper and lower boundary values as $x = 1$ and $x = 0$ so

area of R $= \int_0^1 y dx = \int_0^1 e^{2x+3} dx$

Continued on the next two pages

Step 2: Apply to the equation in step 1 the standard result

$$\int f(ax+b)\mathrm{d}x = \frac{1}{a}g(ax+b)+c$$

where $\frac{\mathrm{d}g}{\mathrm{d}x} = f(ax+b)$ and a, b and c are constant.

Hence $f(ax+b) = e^{2x+3}$ therefore $a = 2$, $b = 3$ and we can ignore c as this is a definite integral.

$$\text{Therefore area of R} = \int_0^1 e^{2x+3}\,\mathrm{d}x$$

$$= \left[\frac{1}{2}e^{2x+3}\right]_0^1 = \left[\frac{1}{2}e^5\right] - \left[\frac{1}{2}e^3\right]$$

$$= \frac{e^5 - e^3}{2} \text{ square units}$$

b) **Step 1:** Use the standard result for finding the volume of a solid of revolution when the region R is rotated through 360° about the x-axis

$$\text{so volume} = \int_0^1 \pi y^2\mathrm{d}x = \int_0^1 \pi(e^{2x+3})^2\mathrm{d}x = \pi\int_0^1 e^{4x+6}\mathrm{d}x$$

Step 2: Apply to the equation in step 1 the standard result

$$\int f(ax+b)\mathrm{d}x = \frac{1}{a}g(ax+b)+c$$

where $\frac{\mathrm{d}g}{\mathrm{d}x} = f(ax+b)$ and a, b and c are constants.

Hence $f(ax+b) = e^{4x+6}$ therefore $a = 4$, $b = 6$ and we can ignore c as this is a definite integral

$$\text{therefore volume} = \pi\int_0^1 e^{4x+6}\mathrm{d}x$$

$$= \left[\frac{\pi}{4}e^{4x+6}\right]_0^1 = \left[\frac{\pi}{4}e^{10}\right] - \left[\frac{\pi}{4}e^6\right]$$

$$= \frac{\pi}{4}(e^{10} - e^6) \text{ cubic units}$$

Method notes

$(e^{(2x+3)})^2 = e^{2(2x+3)} = e^{4x+6}$ using the rule of indices $(e^x)^k = e^{kx}$ where k is a constant.

Finding the volume of a solid of revolution formed by rotating a curve about the y-axis

By a similar argument you can find the volume of a solid of revolution found by rotating the region between a curve, the y-axis and the lines $y = c$ and $y = d$ through 360° about the y-axis.

$$\text{The formula is therefore: volume of revolution} = \int_c^d \pi x^2\mathrm{d}y$$

Exam tips

This result will be in the formula booklet which is provided in the examination. You will not be asked to prove the result but you must learn how to apply the formula.

The solid of revolution is shown in Figure 5.4 below:

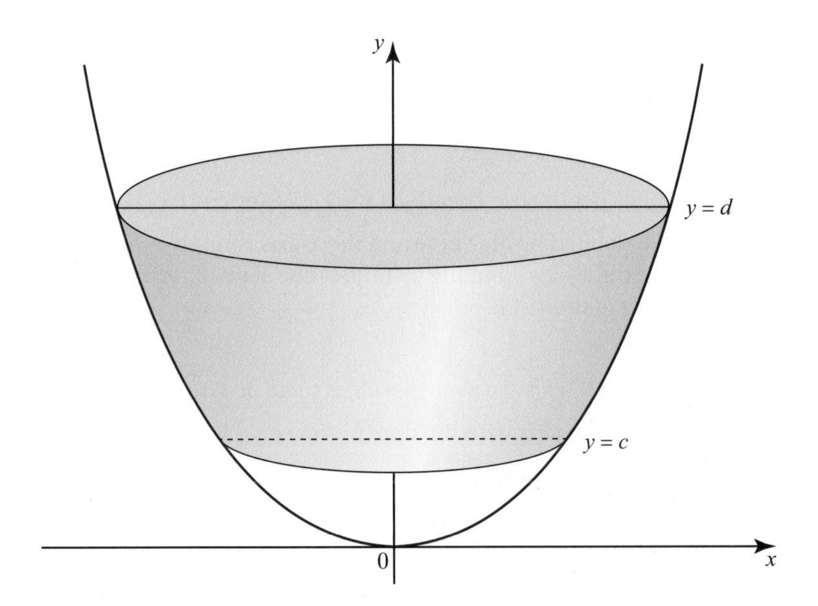

Fig 5.4 The volume of revolution formed by rotating the curve $y = f(x)$ through 360° about the y-axis.

Example

A region S enclosed by the y-axis and the portion of the curve $y = \sqrt{x}$ between $x = 0$ and $x = 4$ is revolved through 360° about the y-axis. Find the volume of the solid of revolution formed by this rotation of S.

Answer

Step 1: If $y = \sqrt{x}$ when $x = 0$ then $y = 0$ and when $x = 4$, $y = \sqrt{4} = 2$.

Step 2: The region S is rotated about the y- axis so apply the formula using the y limits found in Step 1. The upper limit is $y = 2$ and the lower limit is $y = 0$:

volume $= \int_0^2 \pi x^2 \, dy$ so write x in terms of y. The curve is given

by $y = \sqrt{x}$ so squaring both sides gives $x = y^2$.

Step 3: Substitute $x = y^2$ into the formula from step 2:

$$\text{volume} = \int_0^2 \pi x^2 \, dy$$

$$= \int_0^2 \pi (y^2)^2 \, dy$$

$$= \pi \int_0^2 y^4 \, dy \text{ since } \pi \text{ is a constant, therefore:}$$

$$\text{volume} = \left[\frac{\pi}{5} y^5\right]_0^2 = \left[\frac{\pi}{5} 2^5\right] - [0] = \frac{32\pi}{5} \text{ cubic units}$$

Finding the volume of a solid of revolution using parametric equations

In Chapter 3 we introduced parametric equations as an alternative method of defining a curve. You can find volumes of solids of revolution for curves defined in this way as the following example illustrates.

Example

The curve C has parametric equations $x = 1 + 2t$ and $y = t^2 + 4$, $0 \le t \le 1$

The region, R, defined by the curve and the x-axis is rotated through 2π radians about the x-axis. Find the volume of the solid of revolution formed by this rotation of R.

Answer

Step 1: The region R is rotated about the x axis so we need to use the formula volume $= \int \pi y^2 \mathrm{d}x$ with the upper limit of $t = 1$ and the lower limit of $t = 0$ and rewrite in terms of t. Using the chain rule:

$$\text{volume} = \int \pi y^2 \mathrm{d}x = \pi \int y^2 \frac{\mathrm{d}x}{\mathrm{d}t} \mathrm{d}t$$

Step 2: Use the parametric equations $x = 1 + 2t$ so $\frac{\mathrm{d}x}{\mathrm{d}t} = 2$ and $y = t^2 + 4$ and substitute into the equation in step 1:

$$\text{volume} = \pi \int y^2 \frac{\mathrm{d}x}{\mathrm{d}t} \mathrm{d}t = \pi \int_0^1 (t^2 + 4)^2 \times (2)\mathrm{d}t$$

$$= \pi \int_0^1 2(t^4 + 8t^2 + 16)\mathrm{d}t$$

$$= 2\pi \left[\frac{t^5}{5} + 8\frac{t^3}{3} + 16t \right]_0^1$$

$$= 2\pi \left[\frac{1}{5} + \frac{8}{3} + 16 \right] - 2\pi[0]$$

$$= \frac{566\pi}{15} \text{ cubic units}$$

Essential notes

Use of the chain rule to rewrite $\int \mathrm{d}x = \int \frac{\mathrm{d}x}{\mathrm{d}t} \mathrm{d}t$ was covered in Chapter 3 of this book.

Techniques of integration

As discussed previously in this chapter, some integrals have an integrand that is a recognisable function such as

$\int \sin x \mathrm{d}x$, $\int \mathrm{e}^{2x} \mathrm{d}x$ and $\int \sec^2 x \mathrm{d}x$.

For others such as $\int \sin(3x + 4)\mathrm{d}x$ we have discussed the technique of rewriting the integral so that the integrand is transformed into a recognisable function and then the chain rule is applied to obtain the answer for the integral.

Stop and think 2

Find a) $\int \cos x\,dx$ *b)* $\int e^{4x}\,dx$ *c)* $\int \sec x \tan x\,dx$ *d)* $\int \cos(2x-4)\,dx$

There are many others for which an exact integral cannot be found. In these cases we use an approximate or numerical approach such as the trapezium rule which was covered in Core 2

We now introduce methods of integration which transform non-standard functions into standard functions that can be integrated.

Integration by direct substitution for composite functions

The non-standard functions $(2x+5)^4$, $\sin(3x+4)$ are examples of **composite functions** or **functions of a function**. To integrate functions of this type we have to transform them into standard functions as explained in the following example.

Example

Find $\int (2x+5)^4\,dx$

Answer

$(2x+5)^4$ is a non-standard integrand.

Step 1: Choose a new variable u for the function $(2x+5)$

so let $u=(2x+5)$ then $\frac{du}{dx}=2$ therefore $\frac{dx}{du}=\frac{1}{2}$

Step 2: Use the chain rule to rewrite $dx=\frac{dx}{du}du$

Step 3: Substitute the equations from steps 1 and 2 into $\int (2x+5)^4dx$ and integrate with respect to u:

$$\int (2x+5)^4dx=\int u^4\frac{dx}{du}du=\int u^4\frac{1}{2}du=\frac{1}{2}\times\frac{u^5}{5}+c=\frac{1}{10}u^5+c$$

Application of the chain rule

Standard function. No x's allowed in here.

Step 4: Rewrite the answer from step 3 in terms of x using the equation $u=(2x+5)$ from step 1:

$$\frac{1}{10}u^5+c=\frac{1}{10}(2x+5)^5+c$$

Therefore $\int (2x+5)^4dx=\frac{1}{10}(2x+5)^5+c$

Method notes

$(2x+5)^4$ is the composite function $h(g(x))$ where $g: x \to 2x+5$ and $h: x \to x^4$ as discussed in Core 3

u is chosen as $g(x)$ in such composite functions.

Essential notes

Use of the chain rule to rewrite dx by $\frac{dx}{du}du$ was explained in Chapter 3 of this book.

$\int u^4\frac{1}{2}du=\frac{1}{2}\times\frac{u^5}{5}+c$ since 'du' means integrate with respect to u.

Example

Evaluate the following

a) $\int (3x - 8)^2 dx$

b) $\int (x^3 - 5)x^2 dx$

Answer

These are non-standard integrals so the method is to follow the steps as explained in the example above.

a) $\int (3x - 8)^2 dx$

Let $u = 3x - 8$ then $\frac{du}{dx} = 3$ so $dx = \frac{1}{3}du$

Using the chain rule $dx = \frac{dx}{du}du$

Rewriting the integral in terms of u only and then integrating

gives $\int (3x - 8)^2 dx = \int u^2 \times \frac{1}{3} du = \frac{1}{3}\int u^2 du = \frac{1}{9}u^3 + c$

Replacing u in terms of x gives $\int (3x - 8)^2 dx = \frac{1}{9}(3x - 8)^3 + c$

b) $\int (x^3 - 5)x^2 dx$

Let $u = x^3 - 5$ then $\frac{du}{dx} = 3x^2$ so $dx = \frac{1}{3x^2} du$

Using the chain rule $dx = \frac{dx}{du}du$

Rewriting the integral in terms of u only and integrating

gives $\int (x^3 - 5)x^2 dx = \int u \times x^2 \times \frac{1}{3x^2} du = \int \frac{1}{3} u du = \frac{1}{6}u^2 + c$

Replacing u in terms of x gives $\int (x^3 - 5)x^2 dx = \frac{1}{6}(x^3 - 5)^2 + c$

Integration by substitution for product of functions

To integrate functions such as $\sin^2 x \cos x$ and $(2x + 1)e^{x^2+x+3}$ we also transform them into standard functions by a method of substitution. It is important to recognise that these are examples of **products** of two functions of x.

The first function of the integrand product is a composite function.

The second function of the integrand product is the derivative of 'part' of the composite function!

For example $\sin^2 x \cos x$ can be written as $(\sin x)^2 \cos x$.

In this case the first function of the integrand product is the composite function $(\sin x)^2$.

Using conventional labelling we write this as hg(x) where g: $x \rightarrow \sin x$ and h: $x \rightarrow x^2$

The second function of the integrand product is $\cos x$ which is the derivative g'(x) since g(x) = sin x in the composite function.

The following example explains how to use the method of substitution for integrating a product of two such functions.

Example

Evaluate $\int \sin^2 x \cos x dx$

Answer

Step 1: Rewrite the integrand $\sin^2 x \cos x = (\sin x)^2 \cos x$ as the product of two functions hg(x) and g'(x), hg(x) = $(\sin x)^2$, g(x) = sin x and g'(x) = cos x and h: $x \rightarrow x^2$

so $\int \sin^2 x \cos x dx = \int hg(x)\, g'(x)\, dx$

Step 2: Let $u = \sin x$ so $\frac{du}{dx} = \cos x$ and $\frac{dx}{du} = \frac{1}{\cos x}$

Step 3: Use the chain rule to rewrite $dx = \frac{dx}{du} du$

Step 4: Substitute from steps 1 and 2 and 3 into $\int \sin^2 x \cos x dx$, simplify in terms of u only and integrate

so $\int \sin^2 x \cos x dx = \int u^2 \times \cos x \times \frac{1}{\cos x} du = \int u^2 du = \frac{u^3}{3} + c$

Therefore $\int \sin^2 x \cos x dx = \frac{1}{3} \sin^3 x + c$

Method notes

$\int u^2 \times \cos x \times \frac{1}{\cos x} du$ simplifies to $\int u^2 du$ by cancelling $\cos x$. $\int u^2 du$ is then evaluated using the rule of integration.

Example

Evaluate $\int (2x + 1)e^{x^2+x+3} dx$

Answer

This integrand is of a product of two functions of x which can be written as $\int hg(x)\, g'(x) dx$. To evaluate follow the steps explained in the example above.

Step 1: Rewrite $\int (2x + 1)e^{x^2+x+3} dx$ as $= \int hg(x)\, g'(x) dx$ where

g(x) = $x^2 + x + 3$ so g'(x) = $2x + 1$ and h: $x \rightarrow e^x$

Method notes

$\int e^u \times (2x + 1) \times \frac{1}{(2x + 1)} du$ simplifies to $\int e^u du$ by cancelling $(2x + 1)$. $\int e^u du$ is then evaluated using the rule of integration for exponential functions.

Step 2: Let $u = x^2 + x + 3$ so $\frac{du}{dx} = 2x + 1$ and $dx = \frac{1}{(2x + 1)}du$

Step 3: Use the chain rule to rewrite $dx = \frac{dx}{du}du$

Step 4: Substitute from steps 1, 2 and 3 into $\int (2x + 1)e^{x^2+x+3}dx$, simplify in terms of u only and integrate so

$$\int (2x + 1)e^{x^2+x+3}dx = \int e^u \times (2x + 1) \times \frac{1}{(2x + 1)}du$$
$$= \int e^u du = e^u + c$$

Therefore $\int (2x + 1)e^{x^2+x+3}dx = e^{x^2+x+3} + c$

Integrating by substitution for quotient of functions

The method of substitution can also be used to evaluate integrals where the integrand is a quotient of the form $\frac{g'(x)}{g(x)}$ as explained in the following example.

Example

Evaluate the following $\int \frac{2x + 1}{x^2 + x - 3}dx$

Answer

Step 1: Rewrite $\int \frac{2x + 1}{x^2 + x - 3}dx$ as $\int \frac{g'(x)}{g(x)}dx$

where $g(x) = x^2 + x - 3$ and $g'(x) = 2x + 1$

Step 2: Let $u = x^2 + x - 3$ so $\frac{du}{dx} = 2x + 1$and $dx = \frac{1}{2x + 1}du$

Step 3: Use the chain rule to rewrite $dx = \frac{dx}{du}du$

Step 4: Substitute from steps 1, 2 and 3 into $\int \frac{2x + 1}{x^2 + x - 3}dx$, simplify in terms of u only and integrate so

$$\int \frac{2x+1}{x^2+x-3}\,dx = \int \frac{(2x+1)}{u} \times \frac{1}{(2x+1)}\,du$$

$$= \int \frac{1}{u}\,du = \ln|u| + c$$

Therefore $\int \frac{2x+1}{x^2+x-3}\,dx = \ln|x^2+x-3| + c$

Method notes

$\int \frac{(2x+1)}{u} \times \frac{1}{(2x+1)}\,du$ simplifies to $\int \frac{1}{u}\,du$ which is then evaluated as a standard integral.

We can also use the method explained above for quotient integrands of the type $\frac{1}{a}\frac{g'(x)}{g(x)}$ where a is a constant as the next example illustrates.

Example

Evaluate $\int \frac{e^{-2x}}{1+e^{-2x}}\,dx$

Answer

This integrand is a quotient of the type $\frac{1}{a} \times \frac{g'(x)}{g(x)}$ where a is a constant because if $g(x) = 1 + e^{-2x}$ then $g'(x) = -2e^{-2x}$ so the quotient is

$\frac{1}{-2} \times \frac{-2e^{-2x}}{1+e^{-2x}}$ and $a = -2$

To evaluate follow the steps explained in the previous example.

$$\int \frac{e^{-2x}}{1+e^{-2x}}\,dx$$

Let $u = 1 + e^{-2x}$ so $\frac{du}{dx} = -2e^{-2x}$

Using the chain rule $dx = \frac{dx}{du}\,du = \frac{-1}{2e^{-2x}}\,du$

Rewriting $\int \frac{e^{-2x}}{1+e^{-2x}}\,dx$ in terms of u only and then integrating gives

$$\int \frac{e^{-2x}}{1+e^{-2x}}\,dx = \int \frac{e^{-2x}}{u} \times \frac{-1}{2e^{-2x}}\,du = \int -\frac{1}{2u}\,du = -\frac{1}{2}\ln|u| + c$$

Replacing u in terms of x gives $\int \frac{e^{-2x}}{1+e^{-2x}}\,dx = -\frac{1}{2}\ln|1+e^{-2x}| + c$

When evaluating definite integrals using the technique of integration by substitution it is important to change the limits as well as changing the variable in the integrand.

Example

Evaluate $\displaystyle\int_2^5 \frac{x}{(x^2-3)^{\frac{3}{2}}}\mathrm{d}x$

Answer

This integrand is a quotient of the type $\dfrac{1}{a}\dfrac{g'(x)}{g(x)}$ where a is a constant so the method is to follow the steps as explained above.

$$\int_2^5 \frac{x}{(x^2-3)^{\frac{3}{2}}}\mathrm{d}x$$

Let $u = x^2 - 3$ then $\dfrac{\mathrm{d}u}{\mathrm{d}x} = 2x$

Using the chain rule and $\mathrm{d}x = \dfrac{\mathrm{d}x}{\mathrm{d}u}\mathrm{d}u = \dfrac{1}{2x}\mathrm{d}u$

If we are to rewrite $\displaystyle\int_2^5 \frac{x}{(x^2-3)^{\frac{3}{2}}}\mathrm{d}x$ in terms of u only we must find the upper and lower limits of the integral in terms of u.

When $x = 2$ since $u = x^2 - 3$ then $u = 2^2 - 3 = 1$ which is the lower limit for u.

When $x = 5$ since $u = x^2 - 3$ then $u = 5^2 - 3 = 22$ which is the upper limit for u.

Rewriting $\displaystyle\int_2^5 \frac{x}{(x^2-3)^{\frac{3}{2}}}\mathrm{d}x$ in terms of u only and integrating it

gives $\displaystyle\int_2^5 \frac{x}{(x^2-3)^{\frac{3}{2}}}\mathrm{d}x = \int \frac{x}{u^{\frac{3}{2}}} \times \frac{1}{2x}\mathrm{d}u = \int_1^{22} \frac{1}{2}u^{-\frac{3}{2}} = \left[-u^{-\frac{1}{2}}\right]_1^{22}$

Substituting in the upper and lower limits of u gives

$$\left[-u^{-\frac{1}{2}}\right]_1^{22} = \left[-\frac{1}{\sqrt{22}}\right] - \left[-\frac{1}{\sqrt{1}}\right] = 1 - \frac{1}{\sqrt{22}}$$

Therefore $\displaystyle\int_2^5 \frac{x}{(x^2-3)^{\frac{3}{2}}}\mathrm{d}x = 1 - \frac{1}{\sqrt{22}}$

Method notes

Choose an appropriate new variable u which will reduce the integrand to a basic function.

Use the given limits on x to find the limits of the new variable u.

Example

Find the total area enclosed by the curve $y = \frac{10}{2x + 5}$, the y-axis and the line $x = 2$

Answer

Sketch the graph of the function to check that the area is above the x-axis and therefore the integrand is positive.

This is shown in Figure 5.5 below.

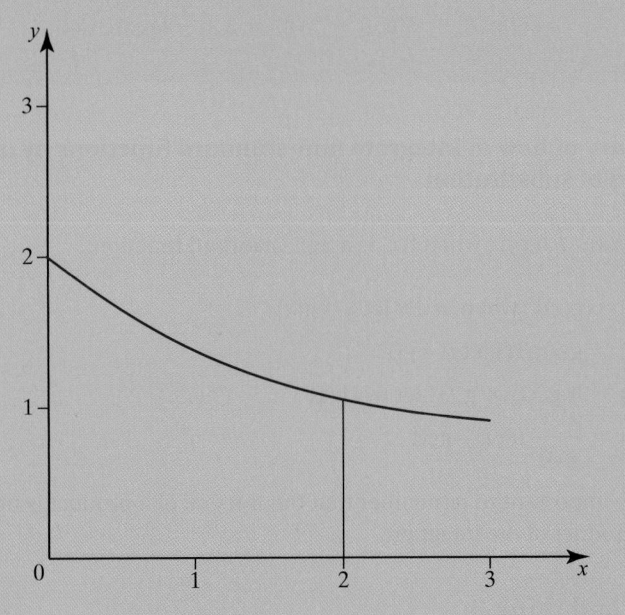

Fig. 5.5 A sketch of the graph of $y = \frac{10}{2x + 5}$ showing the area to be found.

The area A is given by the formula $A = \int_0^2 y \, dx$ where $y = \frac{10}{2x + 5}$

So $A = \int_0^2 \frac{10}{2x + 5} dx$

This integrand is a quotient of two functions of x so it is of the type $\frac{1}{a} \frac{g'(x)}{g(x)}$ where a is a constant and the method is as explained in the last example.

Let $u = 2x + 5$ then $\frac{du}{dx} = 2$

Using the chain rule $dx = \frac{dx}{du} du = \frac{1}{2} du$

Method notes

Finding a finite area A between a curve and the x axis was explained in Core 2

Care must be taken if the area is below the x-axis (negative) as A is then the numerical value of the integral so $A = |\int f(x) \, dx|$.

Continued on the next page

Rewriting the integral in terms of u only means that we must find the upper and lower limit as values of u.

When $x = 0$ since $u = 2x + 5$ so $u = 5$

When $x = 2$ since $u = 2x + 5$ so $u = 9$

$$\text{Therefore } A = \int_0^2 \frac{10}{2x+5}\,dx = \int_5^9 \frac{10}{u} \times \frac{1}{2}\,du = \int_5^9 \frac{5}{u}\,du$$

$$= 5[\ln u]_5^9 = 5[\ln 9 - \ln 5] = 5\ln\frac{9}{5} \text{ square units.}$$

Summary of how to integrate non-standard functions by using a method of substitution

To evaluate $\int f(x)dx$ when $f(x)$ is a non-standard function:

- if $f(x) = (g(x))^n$ where $n \varepsilon \Re$ let $u = g(x)$
- if $f(x) = g(x)g'(x)$ let $u = g(x)$
- if $f(x) = h(g(x)) \times g'(x)$ let $u = g(x)$
- if $f(x) = \frac{g'(x)}{g(x)}$ let $u = g(x)$

It is very important to remember that the integral of a product is not equal to the product of the integrals!

Stop and think 3

By differentiating $f(x) = \frac{x^2e^{4x}}{8} + c$ *where c is a constant show that*

$$\int xe^{4x}dx \neq \frac{x^2}{2} \times \frac{e^{4x}}{4} + c$$

Integration by parts

There are products of functions, such as $\int xe^{4x}dx$ for which integration by substitution does not work.

To develop a rule for the integration of products of functions such as $\int xe^{4x}dx$ we start with the product rule for differentiation which was covered in Core 3

Step 1: State the product rule for the differentiation of two functions of x, which we call u and v:

$$\frac{d(uv)}{dx} = \frac{du}{dx}v + u\frac{dv}{dx}$$

Step 2: Integrate both sides of the equation in step 1 with respect to x so

$$\int \frac{\mathrm{d}(uv)}{\mathrm{d}x}\mathrm{d}x = \int \frac{\mathrm{d}u}{\mathrm{d}x}v\mathrm{d}x + \int u\frac{\mathrm{d}v}{\mathrm{d}x}\mathrm{d}x$$

Step 3: Using the connection between integration and differentiation then $\int \frac{\mathrm{d}(uv)}{\mathrm{d}x}\mathrm{d}x = uv$ so substituting this in the equation from step 2 gives

$$uv = \int \frac{\mathrm{d}u}{\mathrm{d}x}v\mathrm{d}x + \int u\frac{\mathrm{d}v}{\mathrm{d}x}\mathrm{d}x$$

Step 4: Rewriting the equation in step 4 gives

$$\int u\frac{\mathrm{d}v}{\mathrm{d}x}\mathrm{d}x = uv - \int v\frac{\mathrm{d}u}{\mathrm{d}x}\mathrm{d}x$$

This is called the **integration by parts** formula.

Essential notes

The connection between integration and differentiation as an inverse process was discussed in Core 2

Exam tips

You do not need to learn this formula as it will be in your examination booklet but you do need to learn how to apply it.

Using the 'parts' formula for integrating a product of two functions

Example

Find $\int xe^{4x}\,\mathrm{d}x$

Answer

Step 1: State the 'parts' formula $\int u\frac{\mathrm{d}v}{\mathrm{d}x}\mathrm{d}x = uv - \int v\frac{\mathrm{d}u}{\mathrm{d}x}\mathrm{d}x$ and compare the left hand side of this with $\int xe^{4x}\mathrm{d}x$, the integral to be evaluated.

So let $\int u\frac{\mathrm{d}v}{\mathrm{d}x}\mathrm{d}x = \int xe^{4x}\mathrm{d}x$

Step2: Decide on the best choice for u and $\frac{\mathrm{d}v}{\mathrm{d}x}$ from the last equation in step 1 so that when you apply the formula , the integral on the right hand side i.e. $\int v\frac{\mathrm{d}u}{\mathrm{d}x}\mathrm{d}x$ is able to be evaluated.

Therefore let $u = x \Rightarrow \frac{\mathrm{d}u}{\mathrm{d}x} = 1$ and $\frac{\mathrm{d}v}{\mathrm{d}x} = e^{4x} \Rightarrow v = \frac{e^{4x}}{4}$

Step 3: Use the equations from step 2, substitute into the 'parts' formula and integrate so

$$\int xe^{4x}\mathrm{d}x = x \times \frac{e^{4x}}{4} - \int 1 \times \frac{e^{4x}}{4}\mathrm{d}x = \frac{xe^{4x}}{4} - \frac{e^{4x}}{16} + c$$

where c is the constant of integration.

Method notes

Usually you choose to differentiate the power of x which is why $u = x$ in this example. If you choose $\frac{\mathrm{d}v}{\mathrm{d}x} = x$ then $v = \frac{x^2}{2}$ and $u = e^{4x}$ so $\frac{\mathrm{d}u}{\mathrm{d}x} = 4e^{4x}$. Then $\int v\frac{\mathrm{d}u}{\mathrm{d}x}\mathrm{d}x$ becomes $\int \frac{x^2}{2}4e^{4x}\mathrm{d}x$ which is more difficult than the original integral!

Example

Find $\int x^3 \ln x \mathrm{d}x$

Answer

Step 1: State the 'parts' formula $\int u\frac{\mathrm{d}v}{\mathrm{d}x}\mathrm{d}x = uv - \int v\frac{\mathrm{d}u}{\mathrm{d}x}\mathrm{d}x$ and compare the left hand side of this with $\int x^3 \ln x\mathrm{d}x$ the integral to be evaluated.

So let $\int u\frac{\mathrm{d}v}{\mathrm{d}x}\mathrm{d}x = \int x^3 \ln x\mathrm{d}x$

Step 2: Decide on the best choice for u and $\frac{\mathrm{d}v}{\mathrm{d}x}$ from the last equation in step 1 so that when you apply the formula, the integral on the right hand side ie $\int v\frac{\mathrm{d}u}{\mathrm{d}x}\mathrm{d}x$ can be evaluated.

Therefore let $u = \ln x \Rightarrow \frac{\mathrm{d}u}{\mathrm{d}x} = \frac{1}{x}$ and

$$\frac{\mathrm{d}v}{\mathrm{d}x} = x^3 \Rightarrow v = \frac{x^4}{4}$$

Step 3: Use the results from step 2, substitute into the 'parts' formula and integrate

so $\int x^3 \ln x\mathrm{d}x = \frac{x^4}{4}\ln x - \int \frac{x^4}{4} \times \frac{1}{x}\mathrm{d}x = \frac{x^4}{4}\ln x - \int \frac{x^3}{4}\mathrm{d}x$

$= \frac{x^4}{4}\ln x - \frac{x^4}{16} + c$ where c is the constant of integration.

Method notes

In this case you need to choose to integrate the power of x because differentiating $\ln x$ simplifies the integral $\int v\frac{\mathrm{d}u}{\mathrm{d}x}\mathrm{d}x$ in the 'parts' formula. If $\frac{\mathrm{d}v}{\mathrm{d}x} = x^3$ then $v = \frac{x^4}{4}$ and if $u = \ln x$ then $\frac{\mathrm{d}u}{\mathrm{d}x} = \frac{1}{x}$ so $\int v\frac{\mathrm{d}u}{\mathrm{d}x}\mathrm{d}x = \int \frac{x^4}{4} \times \frac{1}{x}\mathrm{d}x = \int \frac{x^3}{4}\mathrm{d}x$

Example

Find $\int x^2 \sin x\mathrm{d}x$

Answer

Step 1: State the 'parts' formula $\int u\frac{\mathrm{d}v}{\mathrm{d}x}\mathrm{d}x = uv - \int v\frac{\mathrm{d}u}{\mathrm{d}x}\mathrm{d}x$ and compare the left hand side of this with $\int x^2 \sin x\mathrm{d}x$ the integral to be evaluated.

So let $\int u\frac{\mathrm{d}v}{\mathrm{d}x}\mathrm{d}x = \int x^2 \sin x\mathrm{d}x$

Step 2: Decide on the best choice for u and $\frac{dv}{dx}$ from the last equation in step 1 so that when you apply the formula, the integral on the right hand side i.e. $\int v\frac{du}{dx}dx$ can be evaluated.

Therefore let $u = x^2 \Rightarrow \frac{du}{dx} = 2x$

and $\frac{dv}{dx} = \sin x \Rightarrow v = -\cos x$

Step 3: Use the results from step 2, substitute into the parts formula and integrate so

$$\int x^2 \sin x dx = -x^2 \cos x - \int 2x(-\cos x)dx = -x^2 \cos x + \int 2x \cos x dx$$

Step 4: Use the parts formula **again** to evaluate $\int 2x \cos x dx$ in step 3 so

for $\int 2x \cos x dx$ let $u = 2x$ and $\frac{dv}{dx} = \cos x$

so $\frac{du}{dx} = 2$ and $v = \sin x$

Step 5: Use the results from step 4, substitute into the parts formula and integrate so $\int 2x \cos x dx = 2x \sin x - \int 2\sin x dx$

$$= 2x \sin x + 2 \cos x + c$$

Step 6: Substitute the answer from step 5 into the equation in step 3 so $\int x^2 \sin x dx = -x^2 \cos x + 2x \sin x + 2\cos x + c$ where c is the constant of integration.

Method notes

In this example you need to use the parts formula twice as the original integrand was $x^2 \sin x$. Whatever the power of x in such a product of functions then that will be the number of times you have to use the parts formula to evaluate the integral.

Stop and think 4

Evaluate $\int_{-1}^{2} (x + 1)e^{-3x}dx$ *giving your answer in terms of e.*

Integration using partial fractions

In Chapter 1 we introduced the method of partial fractions for algebraic fractions with linear and quadratic factors in the denominator.

Differentiating algebraic fractions was often more straightforward when the fraction was expressed as partial fractions.

We now see how partial fractions are useful for integrating algebraic fractions.

Essential notes

Before using partial fractions, check whether substitution might be a quicker method.

Example

a) Write $f(x) = \dfrac{4}{(x-1)(x+3)}$ as the sum of partial fractions.

b) Evaluate $\displaystyle\int \frac{4}{(x-1)(x+3)}\,dx$

Answer

Step 1: Let $\dfrac{4}{(x-1)(x+3)} \equiv \dfrac{A}{x-1} + \dfrac{B}{x+3}$

$$\frac{4}{(x-1)(x+3)} \equiv \frac{A(x+3) + B(x-1)}{(x-1)(x+3)}$$

Step 2: Equate the numerators of the final identity in step 1

so $A(x+3) + B(x-1) = 4$

This identity is true for all values of x so choose suitable numerical values for x in order to evaluate the constants A and B.

Step 3: Let $x = 1$ in the identity in step 2:

$A(4) + B(0) = 4 \Rightarrow A = 1$

Step 4: Let $x = -3$ in the identity in step 2:

$A(0) + B(-4) = 4 \Rightarrow B = -1$

Step 5: Substitute the values for A and B from step 4 into the first identity in step 1 so $f(x) = \dfrac{4}{(x-1)(x+3)} = \dfrac{1}{x-1} - \dfrac{1}{x+3}$

b) Use the answer for $f(x)$ in terms of partial fractions from part a) and then integrate $f(x)$:

$$\int \frac{4}{(x-1)(x+3)}\,dx = \int \left(\frac{1}{x-1} - \frac{1}{x+3}\right) dx$$

$= \ln|x-1| - \ln|x+3| + c$

$= \ln\left|\dfrac{x-1}{x+3}\right| + c$ where c is the constant of integration.

Example

Use partial fractions to find $\displaystyle\int \frac{x}{(x+1)(x+2)^2}\,dx$.

Answer

Step 1: Let $\dfrac{x}{(x+1)(x+2)^2} \equiv \dfrac{A}{x+1} + \dfrac{B}{x+2} + \dfrac{C}{(x+2)^2}$

Then simplify the right hand side of the identity as one fraction

$$\frac{x}{(x+1)(x+2)^2} \equiv \frac{A(x+2)^2 + B(x+1)(x+2) + C(x+1)}{(x+1)(x+2)^2}$$

Method notes

The use of partial fractions for rational functions with a 'repeated' denominator was covered in Core 3

In this example $(x+2)^2$ is the 'repeated' denominator.

Step 2: Equate the numerators of the final identity in step 1:

$x = A(x+2)^2 + B(x+1)(x+2) + C(x+1)$

This identity is true for all values of x so choose suitable values for x to evaluate the constants A, B and C.

Step 3: Let $x = -1$ in the identity in step 2:

$-1 = A(1^2) + B(0)(1) + C(0) \Rightarrow A = -1$

Let $x = -2$ in the identity in step 2:

$-2 = A(0^2) + B(-1)(0) + C(-1) \Rightarrow C = 2$

Let $x = 0$ in the identity in step 2:

$0 = A(2^2) + B(1)(2) + C(1) \Rightarrow B = 1$

Step 4: Substitute the values for A, B and C in the identity in step 1:

$$\frac{x}{(x+1)(x+2)^2} \equiv \frac{-1}{x+1} + \frac{1}{x+2} + \frac{2}{(x+2)^2}$$

Step 5: Integrate with respect to x the identity in step 4:

$$\int \frac{x}{(x+1)(x+2)^2}\,dx = \int \frac{-1}{x+1} + \frac{1}{x+2} + \frac{2}{(x+2)^2}\,dx$$

$$= -\ln|x+1| + \ln|x+2| - \frac{2}{(x+2)} + c$$

$$= \ln\left|\frac{x+2}{x+1}\right| - \frac{2}{(x+2)} + c$$

where c is a constant of integration.

Method notes

Be careful with integrals of the form $\int \frac{A}{(x+a)^2}\,dx$. The answer is not a logarithm! Write $\frac{A}{(x+a)^2}$ as $A(x+a)^{-2}$ and then

$$\int \frac{A}{(x+a)^2}\,dx = \int A(x+a)^{-2}\,dx$$

$$= -A(x+a)^{-1} = -\frac{A}{(x+a)}$$

Example

Find $\int \frac{2x}{(x-1)(x+1)}\,dx$

Answer

Step 1: Multiply out the denominator and check whether the integrand is of the non-standard type $\frac{g'(x)}{g(x)}$

$(x-1)(x+1) = x^2 - 1$ so if $g(x) = x^2 - 1$ then $g'(x) = 2x$

Step 2: Integrate using the result from step1 for this non-standard integrand which is $\frac{g'(x)}{g(x)}$:

$$\int \frac{2x}{(x-1)(x+1)}\,dx = \int \frac{2x}{(x^2-1)}\,dx = \ln|x^2-1| + c$$

where c is the constant of integration.

Method notes

In this example, a substitution is more straight forward than using partial fractions. The numerator is equal to the derivative of the denominator. Let $u = g(x) = (x^2 - 1)$ then $\frac{du}{dx} = g'(x) = 2x$

$$\int \frac{2x}{(x^2-1)}\,dx = \int \frac{1}{u}\,du$$

$$= \ln|u| + c$$

Solving first order differential equations

In Chapter 4 we introduced the idea of a differential equation as an equation in terms of an independent variable, a dependent variable and (some of) its derivatives. For example,

$\frac{dN}{dt} = -kN$ is an example of a **first order differential equation** because it involves first derivatives only. N is the dependent variable and t the independent variable. k is a constant.

Essential notes

The symbol $\frac{dy}{dx}$ is the **first** derivative of y with respect to x.

It means y has been differentiated **once** with respect to x.

To find the solution of the differential equation $\frac{dy}{dx} = (x^2 + x - 1)$ means we have to find y in terms of x . y is the dependent variable and x the independent variable.

The method of solution is to integrate once with respect to x. This is possible since the only variable on the right hand side of the equation is x so each term can be integrated with respect to x.

Essential notes

$$\int \frac{dy}{dx} dx = \int dy = y$$

by using the chain rule.

This was covered in Core 3

$$\int \frac{dy}{dx} dx = \int (x^2 + x - 1)\, dx$$

so $\int \frac{dy}{dx} dx = y = \frac{x^3}{3} + \frac{x^2}{2} - x + c$ where c is the constant of integration.

We now extend the method of solution to first order differential equations of the form

$$\frac{dy}{dx} = f(x)g(y)$$

which means $\frac{dy}{dx}$ is equal to a product of two functions, one of which is a function of x and the other which is a function of y.

If $\frac{dy}{dx} = f(x)g(y)$ then dividing both sides of this equation by $g(y)$ gives

$$\frac{1}{g(y)} \frac{dy}{dx} = f(x).$$

If we then integrate the equation with respect to x this gives

$$\int \frac{1}{g(y)} \frac{dy}{dx} dx = \int \frac{1}{g(y)} dy = \int f(x) dx$$

You can see that the 'variables' have 'separated' so that only the variable y is on the left-hand side of the equation and the integration on this side is with respect to y. The only variable on the right-hand side of the equation is now x and the integration on this side is with respect to x so each side of the equation can then be integrated.

Therefore if $\frac{dy}{dx} = f(x)g(y)$ then $\int \frac{1}{g(y)}dy = \int f(x)dx$

and this is called the method of separation of variables.

Exam tips

You must learn how to use this method of separation of variables as it is likely to be tested in the examination.

Example

Find the general solution of the differential equation $\frac{dy}{dx} = x(y + 1)$

Answer

You should recognise that this is a first order differential equation of the type

$\frac{dy}{dx} = f(x)g(y)$ where $f(x) = x$ and $g(y) = y + 1$ therefore the method of solution will be to separate the variables.

Step 1: Given $\frac{dy}{dx} = x(y + 1)$ to separate the variables, divide both sides by $(y + 1)$:

$$\frac{1}{(y+1)}\frac{dy}{dx} = x$$

Step 2: Integrate both sides of the last equation in step 1 with respect to x:

$$\int \frac{1}{y+1}dy = \int x dx$$

$$\Rightarrow \ln(y+1) = \frac{x^2}{2} + c$$

where c is a constant of integration.

Step 3: Write the equivalent statement of the final equation in step 2:

$y + 1 = e^{\frac{x^2}{2}+c} = e^{\frac{x^2}{2}} \times e^c$ (using rules of indices).

which is the general solution but it can be simplified further.

Step 4: Simplify the equation in step 3:

$y = -1 + e^c e^{\frac{x^2}{2}} = -1 + Ae^{\frac{x^2}{2}}$ where A is the constant e^k

Therefore $y = -1 + Ae^{\frac{x^2}{2}}$ where A is a constant.

Essential notes

Equivalent statements were covered in Core 3

The solution $y = -1 + Ae^{\frac{x^2}{2}}$ is called the **general solution** of the differential equation. It contains one unknown constant A.

The general solution is a **family of solutions** and each value of the constant A is a **particular solution** of the differential equation.

To find a particular value of A we need to be given values of x and y.

For example if we are given that $y = 2$ when $x = 1$ then substituting these values into the general solution $y = -1 + Ae^{\frac{x^2}{2}}$ gives

$2 = -1 + Ae^{\frac{1^2}{2}} \Rightarrow A = 3e^{-\frac{1}{2}}$.

So $y = -1 + 3e^{-\frac{1}{2}}e^{\frac{x^2}{2}}$ and using the rule of indices to simplify $3e^{-\frac{1}{2}}e^{\frac{x^2}{2}}$ gives

$y = -1 + 3e^{\frac{x^2}{2} - \frac{1}{2}}$ which is the **particular** solution of the differential equation $\frac{dy}{dx} = x(y + 1)$.

$y = 2$ when $x = 1$ is called a **boundary condition.**

Example

Given the differential equation $\frac{dy}{dx} = \frac{\sin x}{y}$

and that $1 > 2\cos x$, find y^2 as a function of x.

Answer

You should recognise that this is a first order differential equation of the type $\frac{dy}{dx} = f(x)g(y)$ therefore the method of solution will be to separate the variables.

Step 1: Separate the variables and integrate:

$$\frac{dy}{dx} = \frac{\sin x}{y} \text{ so } \int y\,dy = \int \sin x\,dx$$

therefore $\frac{y^2}{2} = -\cos x + c$ where c is a constant of integration.

Step 2: Use the boundary conditions $x = \pi$ when $y = 1$ in the general solution from step 2

$$\text{so } \frac{1^2}{2} = -\cos \pi + c \Rightarrow c = -\frac{1}{2}$$

Step 3: Substitute the value of c into the general solution of step 1 so the particular solution of the first order differential equation

$\frac{dy}{dx} = \frac{\sin x}{y}$ is $\frac{y^2}{2} = -\cos x - \frac{1}{2}$ so $y^2 = -(2\cos x + 1)$

Example

The acceleration of a parachutist is given by $\frac{dv}{dt} = -0.2(v + v^2)$ where v is the speed of the parachutist in m s^{-1} at any given time t seconds.

The initial value of v is 40 m s^{-1}.

Find the particular solution of v.

Answer

You should recognise that this is a first order differential equation of the type

$\frac{dv}{dt} = f(t)g(v)$ where $f(t) = 0.2 = 0.2\, t^0$ and $g(v) = v + v^2$ therefore the method of solution will be to separate the variables.

Step 1: Separate the variables and integrate so

$$\frac{1}{v + v^2}\frac{dv}{dt} = -0.2 \Rightarrow \int \frac{1}{v + v^2} dv = \int -0.2 dt$$

Step 2: Use partial fractions to simplify the non-standard integrand $\frac{1}{v + v^2}$

$$\text{so } \frac{1}{v + v^2} = \frac{1}{v(1 + v)} = \frac{A}{v} + \frac{B}{1 + v} = \frac{A(1 + v) + Bv}{v(1 + v)}$$

Step 3: Equate the numerators of the final identity in step 2

so $A(1 + v) + Bv = 1$

This identity is true for all values of v so choose suitable values of v to evaluate the constants A and B.

Step 4: Let $v = 0$, in the identity in step3 $\Rightarrow A = 1$

Let $v = -1$ in the identity in step 3 $\Rightarrow B = -1$

Step 5: Substitute the values of A and B into the identity in step 2 and integrate

$$\text{so } \int \frac{1}{v + v^2} dv = \int \frac{1}{v(1 + v)} dv = \int \left(\frac{1}{v} - \frac{1}{1 + v}\right) dv$$

$$= \ln v - \ln(1 + v) = \ln \frac{v}{1 + v}$$

Essential notes

If the independent variable is t and a condition is given when $t = 0$ the boundary condition is usually called **an initial condition**.

$v = 40$ when $t = 0$ is an initial condition.

Continued on the next two pages

Step 6: Evaluate the right hand side of the equation in step 1

so $\int -0.2\text{d}t = -0.2t$

Step 7: Complete the integration from the equation in step 1 by equating the answers from steps 5 and 6

so $\ln\frac{v}{1+v} = -0.2t + c$ where c is a constant of integration.

Step 8: Write the equivalent statement of the equation in step 7 so

$\frac{v}{1+v} = e^{-0.2t+c} = Ce^{-0.2t}$ where $C = e^c$ and is therefore a constant.

Step 9: Use the boundary condition that when $t = 0$, $v = 40$ and substitute into the equation of step 8 $\Rightarrow C = \frac{40}{41}$

Step 10: Substitute the value of C into the equation in step 8 so

$$\frac{v}{1+v} = \frac{40}{41}e^{-0.2t}$$

Step 11: Simplify the algebra of the equation in step 10 to make v the subject of the equation so $v = (1+v)\frac{40}{41}e^{-0.2t}$

$$\Rightarrow v = \frac{40}{41}ve^{-0.2t} + \frac{40}{41}e^{-0.2t}$$

So $v - \frac{40}{41}ve^{-0.2t} = \frac{40}{41}e^{-0.2t} \Rightarrow \frac{1}{41}v(41 - 40e^{-0.2t}) = \frac{40}{41}e^{-0.2t}$

Therefore $v = \frac{40e^{-0.2t}}{41 - 40e^{-0.2t}}$ is the particular solution of v for the given boundary conditions.

Numerical integration

When an integral can be evaluated to give an exact solution, the answer is called an **analytical solution** of the integral.

However, there are many integrals for which an analytical solution cannot be found.

For example, $\int_0^2 e^{-x^2}\text{d}x$ does not have an analytical solution.

To evaluate integrals of this kind we use a numerical method such as the trapezium rule that you met in Core 2 and the solution is called a **numerical solution**.

The trapezium rule as a general formula for approximating an integral is

$$\int_a^b y\,dx \approx \frac{1}{2}h[y_0 + 2(y_1 + y_2 + y_3 + \ldots + y_{n-2} + y_{n-1}) + y_n]$$

where $h = \dfrac{b - a}{n}$ and $y_i = f(a + ih)$

The following example revises how to use the trapezium rule in the context of finding a numerical solution to an integral which does not have an analytical solution.

Example

Use the trapezium rule with eight strips to estimate the value of

$\int_0^2 e^{-x^2}\,dx$ correct to 2 decimal places.

Answer

Step 1: Find the value of h the width of each strip given $a = 0$ $b = 2$ and

$n = 8$ as there are eight strips so $h = \dfrac{2 - 0}{8} = \dfrac{1}{4} = 0.25$

Step 2: Evaluate y_i and complete the table of values for x and y with $h = 0.25$ so

x	0	0.25	0.5	0.75	1	1.25	1.5	1.75	2
y	1	0.9394	0.7788	0.5698	0.3679	0.2096	0.1054	0.0468	0.0183

Step 3: Apply the trapezium rule formula so $\int_0^2 e^{-x^2}\,dx$

$$\approx \frac{1}{2} \times \frac{1}{4}[1 + 2(0.9394 + 0.7788 + 0.5698 + 0.3679 + 0.2096 + 0.1054 + 0.0468) + 0.0183] = 0.8817$$

$\int_0^2 e^{-x^2}\,dx = 0.88$ correct to 2 decimal places.

Stop and think answers

1. *To work out* $\int \sin^2 x dx$ *we must rewrite using the double angle formula* $\cos 2x = 1 - 2\sin^2 x$ *and* $\sin^2 x = \frac{1}{2}(1 - \cos 2x)$

   Therefore $\int \sin^2 x dx = \int \frac{1}{2}(1 - \cos 2x)\, dx = \frac{x}{2} - \frac{\sin 2x}{4} + c$ *where* c *is a constant of integration.*

2. *You should recognise all these integrals as standard integrals and therefore you can just state the answer:*

   a) $\int \cos x dx = \sin x + c$

   b) $\int e^{4x} dx = \frac{e^{4x}}{4} + c$

   c) $\int \sec x \tan x dx = \sec x + c$

   d) $\int \cos(2x - 4)\, dx = \frac{\sin(2x - 4)}{2} + c$

3. *So if* $\int xe^{4x} dx$ *is equal to* $\frac{x^2 e^{4x}}{8} + c$ *then when* $\frac{x^2 e^{4x}}{8} + c$ *is differentiated it should give* xe^{4x}

   To differentiate $\frac{x^2 e^{4x}}{8}$ *with respect to* x *you should recognise that it is a product of two functions of* x *so use the product rule:*

   Let $u = \frac{x^2}{8} \Rightarrow \frac{du}{dx} = \frac{x}{4}$ *and* $v = e^{4x} \Rightarrow \frac{dv}{dx} = 4e^{4x}$ *so the product rule gives the differentiation of* $\frac{x^2 e^{4x}}{8} + c = \frac{x^2}{8} \times 4e^{4x} + e^{4x} \times \frac{x}{4}$

   $= \frac{x^2 e^{4x}}{2} + \frac{xe^{4x}}{4}$ *which is not* xe^{4x} *so we have shown that* $\int xe^{4x} dx \neq \frac{x^2 e^{4x}}{8} + c.$

4. *To evaluate* $\int_{-1}^{2} (x+1)e^{-3x}\,dx$ *use the parts formula with* $u = x + 1$

$$\Rightarrow \frac{du}{dx} = 1 \text{ and } \frac{dv}{dx} = e^{-3x} \Rightarrow v = \frac{-e^{-3x}}{3}$$

So
$$\int_{-1}^{2} (x+1)e^{-3x}\,dx = \left[\frac{-e^{-3x}}{3}(x+1)\right] - \int \frac{-e^{-3x}}{3}\,dx$$

$$= \left[\frac{-e^{-3x}}{3}(x+1) - \frac{e^{-3x}}{9}\right]_{-1}^{2}$$

$$= -e^{-6} - 0 - \frac{1}{9}[e^{-6} - e^{3}]$$

$$= -\frac{10}{9}e^{-6} + \frac{e^{3}}{9}$$

This chapter introduces the idea of a vector and the algebra of vectors. In pure mathematics vectors are important in geometry and allow mathematicians to explore algebraically complex problems in two and three dimensions. Vectors are also important in applied mathematics, science and engineering and can be used to represent any quantity that has both a magnitude and direction, such as forces, velocity and acceleration. You may have come across the use of vectors if you are studying Mechanics as one of your A level modules.

To introduce the ideas in this chapter consider the following example of a 'mathematical journey'.

Example

John and Christine set off for a walk across country.

They set off from their starting point and first walk a distance of 4 km in a north east direction and then
3 km due east and stop for a picnic.

a) How far have they walked from setting out to their picnic site?

b) How far to the east is their final position relative to the initial position?

c) How far to the north is their final position relative to the initial position?

d) In what direction have they finished relative to the original position?

e) What is the 'straight line' distance from the original position to the final position?

Answer

a) **Step 1:** Draw a diagram of the journey as shown in Figure 6.1 below. The starting point is labelled as O. Point A is where they changed direction. Point B is the end point of the journey.

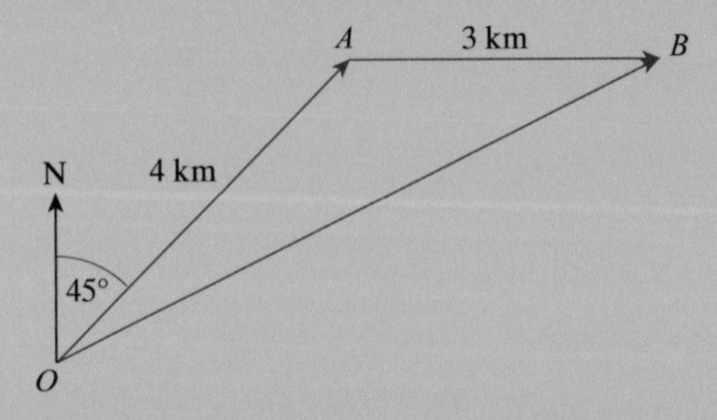

Fig. 6.1
The walk from O to B via A.

Step 2: John and Christine have walked from O to A then A to B so the total distance walked is length OA + length $AB = 4 + 3 = 7$ km.

b) **Step 1:** Draw the line AM where M is the point directly below A and M is at the horizontal level of O. OM is the distance east of their initial position when they are at the point A as shown in Figure 6.2 below. So in triangle OAM,

$$\frac{OM}{4} = \cos 45° \Rightarrow OM = 4\cos 45° = 2\sqrt{2}\text{ km}$$

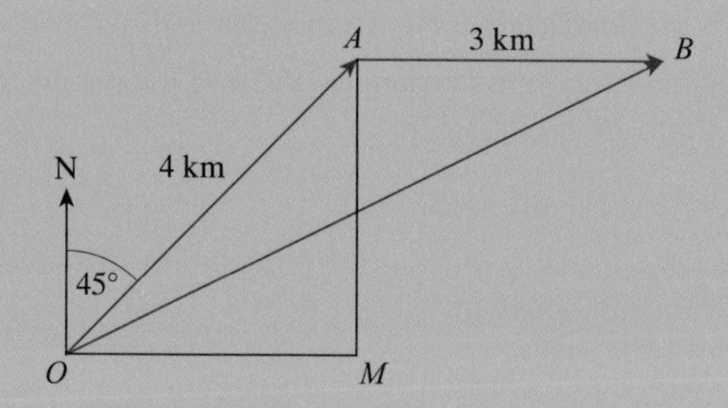

Fig. 6.2
The position east of the final position relative to the initial position.

Step 2: $OM + AB$ is the distance east of the final position, point B, relative to their initial position O. As $AB = 3$ km and using the result in step 2:

$OM + AB = 2\sqrt{2} + 3$ km

c) From Figure 6.2 the length of AM is how far the point A is to the north of O.

Find AM by using trigonometry in triangle OMA:

$$\frac{AM}{4} = \sin 45° \Rightarrow AM = 4\sin 45° = 2\sqrt{2}\text{ km}$$

Therefore the final position, point B, is $2\sqrt{2}$ km to the north of the starting point O.

d) The direction of the final position relative to the starting position is the bearing of B relative to O shown by angle α in the Figure 6.3 below.

Find angle α by using trigonometry in triangle OBC where $\angle OCB = 90°$:

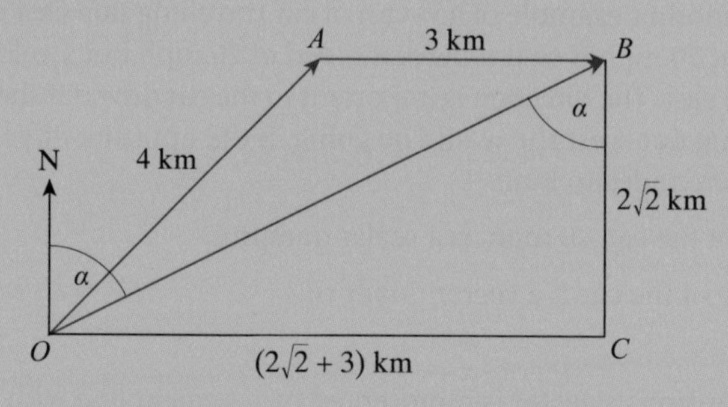

Fig. 6.3
The direction of the final position relative to the initial position.

so $\tan\alpha = \dfrac{OC}{CB}$ and using the answers from b) and c)

$$\tan\alpha = \frac{2\sqrt{2}+3}{2\sqrt{2}} = 2.061 \Rightarrow \alpha = 64° \text{ (to nearest degree).}$$

Continued on the next page

Essential notes

Bearings were covered in your GCSE course. The 'bearing' angle is measured clockwise from North and the distance is measured from the 'relative point' which in this case is O.

Therefore the final position has a bearing of 064° or N 64° E relative to the original position O so they finished in a direction N 64° E relative to their original position.

e) The 'straight line' distance from the original position to the final position is the length OB as shown in Figure 6.3.

Use Pythagoras' theorem in triangle OBC and the results from parts b) and c) then $OB^2 = OC^2 + CB^2$:

$$OB^2 = (2\sqrt{2} + 3)^2 + (2\sqrt{2})^2$$

Therefore $OB = 6.48$ km which is the straight line distance from the original to the final position.

The journey from O to B consists of two parts: the journey from O to A followed by the journey from A to B.

Each displacement has a size (or magnitude) and a direction. For example, the displacement from O to A has size 4 km and direction 'north-east' (or a bearing of 045°).

Quantities such as the 'journeys' described in the previous example which have size (or magnitude) and direction are called **displacements** and are examples of **vectors**. They are often called **displacement vectors**.

Definitions

A quantity that has just size and **no** direction is called a **scalar** quantity.

A quantity that has both size (called **magnitude**) and direction is called a **vector** quantity.

In the example above the total distance (or length) of the journey, 7 km, is a scalar quantity.

The final displacement from the initial position was 6.48 km on a bearing N 64°E which is an example of a vector quantity.

Velocity is another example of a vector: a car travelling due east on a motorway at 70 mph is said to have a speed of 70 mph but a velocity of 70 mph due east. The direction is important to the car driver! If she happened to be heading due west she would be going in the opposite direction and reach a different destination.

The speed of the car, 70 mph, is a scalar quantity.

The velocity of the car is a vector quantity.

Notation

In two dimensions a vector is represented by a straight line with an arrowhead. This is sometimes called a **directed line segment**. To describe the direction of a vector clearly we usually state the angle made by the vector with a fixed line such as the x-axis, y-axis or with the direction due North.

Fig. 6.4 shows a displacement vector from P to Q represented by **p** or $\overrightarrow{PQ}$. On the diagram the vector is labelled $\overrightarrow{PQ}$ or **p**.

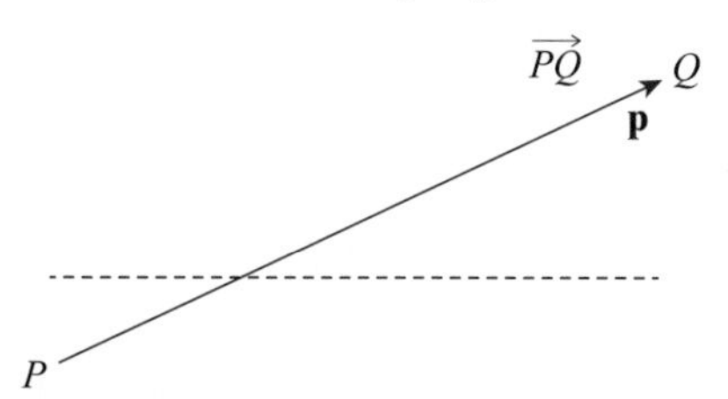

Fig. 6.4
Displacement vector from P to Q.

In this book a vector is printed as a bold letter for example, **p** or $\overrightarrow{PQ}$.

When you write vectors by hand it is usual to underline the symbol representing the vector, for example p or PQ as this is easier than writing in bold!

The magnitude of the vector is then written as p or PQ (not underlined). It can also be written as |p| and is read as 'the modulus of the vector **p**'. This is a scalar quantity.

In the example above, the displacement from O to A is written as $\overrightarrow{OA}$ and the displacement from A to B is written as $\overrightarrow{AB}$ and both are vector quantities.

Essential notes

The modulus symbol was covered in Core 3 and it means the numerical value of a quantity in this case the magnitude only of the vector **p**.

The triangle law of vector addition

Suppose that in the previous example, a third person, Sue, had decided to walk in a straight line from O to B to meet up with John and Christine. The three walkers started and finished in the same place but their journeys were different. Therefore as was calculated previously, Sue's displacement vector would have magnitude 6.48 km and a direction of N 64° E.

So comparing the two journeys:

For John and Christine it was O to A followed by A to B which in vector notation is written as

$\overrightarrow{OA} + \overrightarrow{AB}$.

For Sue it was from O directly to B which in vector notation is $\overrightarrow{OB}$

Therefore we can write $\overrightarrow{OA} + \overrightarrow{AB} = \overrightarrow{OB}$.

This is called the '**triangle law**' of vector addition and it is illustrated diagrammatically in Figure 6.5 below.

Essential notes

The addition symbol + written between these two vectors is translated as 'followed by' because the end point A of the vector $\overrightarrow{OA}$ is the beginning point A of the vector $\overrightarrow{AB}$.

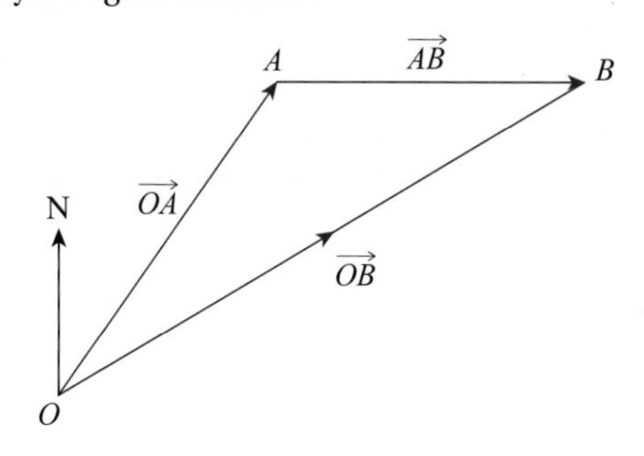

Fig. 6.5
The 'triangle law' of vector addition.

Cartesian components of a vector

The displacement from O to B ($\overrightarrow{OB}$) can also be described as a displacement of $(2\sqrt{2} + 3)$ km in a direction due east (O to C in Figure 6.3) followed by a displacement of $2\sqrt{2}$ km in a direction due north (C to B in Figure 6.3) as was calculated in parts b) and c) respectively of the example on page 91. In parts d) and e) this was also shown to be a displacement vector having magnitude 6.48 km in a direction of N 64° E.

The perpendicular distances in the example above, $(2\sqrt{2} + 3)$ km in a direction due east and $2\sqrt{2}$ km in a direction due north, are called the **components** of the displacement vector $\overrightarrow{OB}$. Components of vectors provide an alternative way of describing a vector giving a link between a geometric approach and an algebraic approach to problem solving with vectors.

Cartesian coordinates use the perpendicular axes x and y through an origin O. We use **Cartesian components** to describe vectors, which are given in terms of two specified directions that are perpendicular to each other and in two dimensions these are the x- and y-axes. The following example illustrates the simplicity of vectors expressed in component form.

Example

A vector **a** has magnitude 9 units and direction 50°. Write **a** in component form.

Answer

Step 1: Let $\mathbf{a} = \overrightarrow{AB}$ and draw a diagram showing the magnitude and direction starting from the point A with Cartesian axes through the point A as shown in Figure 6.6 below.

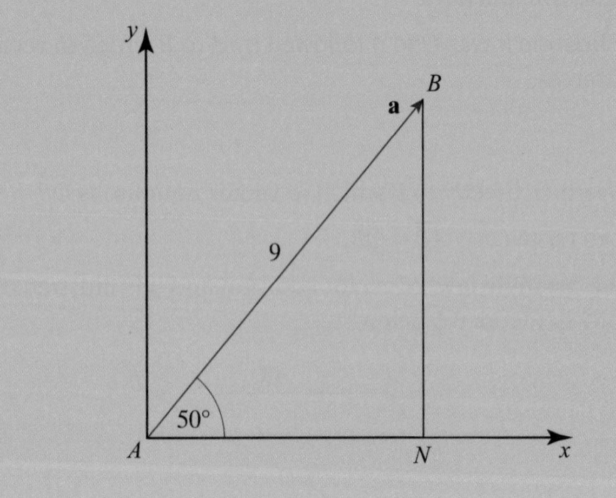

Fig. 6.6
The vector **a** is $\overrightarrow{AB}$ in the triangle ANB.

Step 2: Find AN using trigonometry in the right angled triangle ANB:

$$\frac{AN}{9} = \cos 50° \Rightarrow AN = 9 \cos 50° = 5.79$$

Therefore the 'x' component of $\overrightarrow{AB}$ (or component in the positive x direction) is 5.79

Step 3: Find NB using trigonometry in triangle ANB:

$$\frac{NB}{9} = \sin 50° \Rightarrow NB = 9 \sin 50° = 6.89$$

So the 'y' component of $\overrightarrow{AB}$ (or component in the positive y direction) is 6.89

Cartesian component notation (or column vector notation)

From the last example the vector **a** is written in **Cartesian component form** as $\mathbf{a} = \begin{pmatrix} 5.79 \\ 6.89 \end{pmatrix}$ and this is also called **column vector form.** The 'top' entry in the column is the 'x component 'and the bottom entry is the 'y component'.

Example

A vector **b** has magnitude 12 units and direction 210°. Write **b** in component form.

Answer

Step 1: Let $\mathbf{b} = \overrightarrow{OB}$ and draw a diagram showing the magnitude and direction starting from the point O with Cartesian axes drawn through O as shown in Figure 6.7 below.

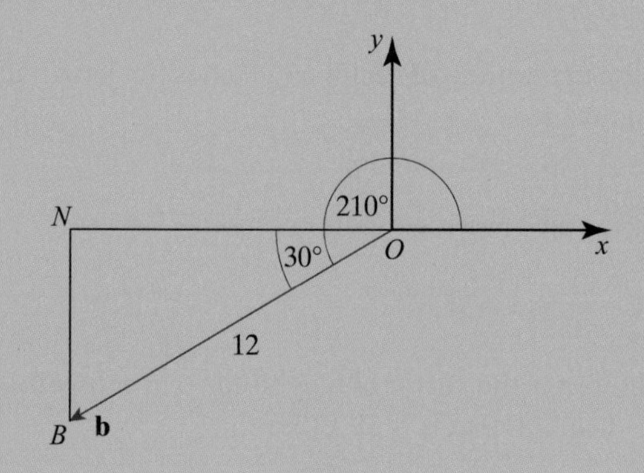

Step 2: Find ON using trigonometry in the right angled triangle OBN:

$\frac{ON}{12} = \cos 30°$ but ON is in the negative direction of the x-axis so

$-ON = -12 \cos 30° = -10.39$

Therefore the x component of the vector $\overrightarrow{OB}$ is -10.39

Essential notes

The direction of a vector is measured from the positive direction of the x-axis. Angles of vectors are measured as positive when rotated anti-clockwise from the positive direction of the x-axis. This was discussed in Core 2.

Fig. 6.7
The vector **b** is $\overrightarrow{OB}$ in the triangle OBN.

Continued on the next page

Step 3: Find NB using trigonometry in triangle OBN:

$\frac{NB}{12} = \cos 60°$ but NB is in the negative direction of the y-axis so

$-NB = -12 \cos 60° = -6$

Therefore in Cartesian component form $\mathbf{b} = \begin{pmatrix} -10.39 \\ -6 \end{pmatrix}$.

Example

A vector $\mathbf{c} = \begin{pmatrix} 4 \\ 3 \end{pmatrix}$ Show the vector **c** on a diagram and find the magnitude and direction of **c**.

Answer

Step 1: Draw a diagram which translates the Cartesian component information for the vector **c** which is 4 units in the x-direction and 3 units in the y-direction as shown in Figure 6.8 below.

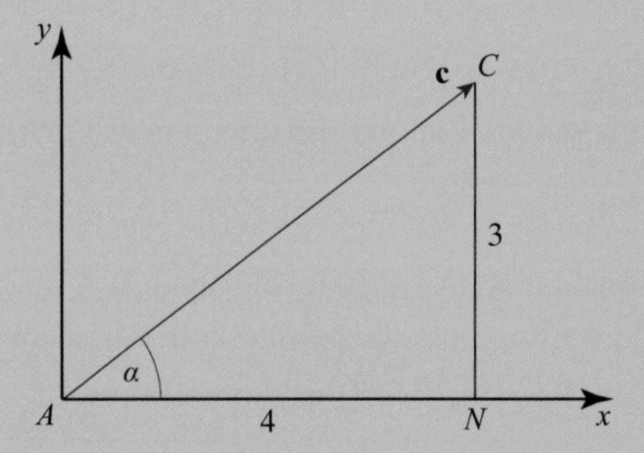

Fig. 6.8
Diagram showing the vector **c** ($\overrightarrow{AC}$).

Step 2: Find the magnitude of vector **c** (which is the length of AC) using Pythagoras so

$AC^2 = 4^2 + 3^2 = 25 \Rightarrow AC = 5$

Step 3: Find the angle α using trigonometry so

$\tan \alpha = \frac{3}{4} \Rightarrow \alpha = 36.87°$

Therefore **c** is the vector of magnitude 5 in a direction N (90° – 36.87°) E that is N 53.13°E.

Example

A vector $\mathbf{d} = \begin{pmatrix} -5 \\ 8 \end{pmatrix}$. Show the vector **d** on a diagram and find the magnitude and direction of **d**.

Answer

Step 1: Draw a diagram which translates the Cartesian component information for the vector **d** which is –5 units in the x-direction and 8 units in the y-direction as shown in Figure 6.9 below.

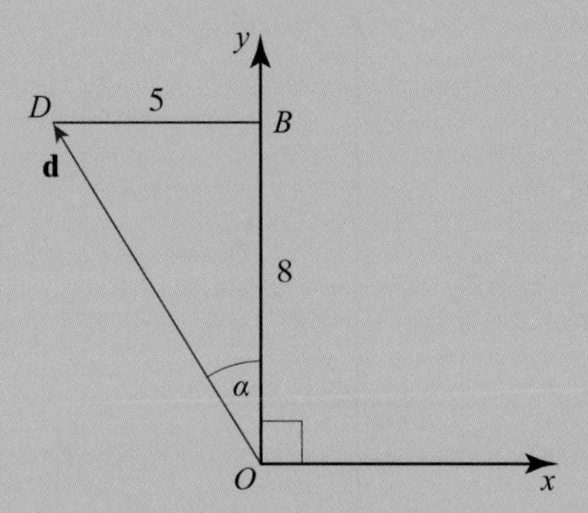

Fig. 6.9
Diagram showing the vector **d** ($\overrightarrow{OD}$).

Step 2: Find the magnitude of vector **d** which is the length of OD using Pythagoras:

$OD^2 = 5^2 + 8^2 = 89 \Rightarrow OD = 9.43$

Step 3: Find the direction of **d** which is the angle $90° + \alpha°$ (direction is given from the positive x-axis) by using trigonometry in the triangle:

$$\tan \alpha = \frac{5}{8} \Rightarrow \alpha = 32.01°$$

Therefore the magnitude and direction of **d** are 9.43 and 122.01° respectively.

Essential notes

It usually helps to draw a diagram showing the direction of the vector. Put the distances on the diagram and remember that distances are positive numbers!

Algebra of vectors

Equal vectors

Two vectors are equal if

- they have the same magnitude, and
- they point in the same direction.

It is important to note that it doesn't matter where on the page a vector is drawn. All the vectors **a** in the Figure 6.10 are equal because they are the same length and point in the same direction. These are often called **free vectors**.

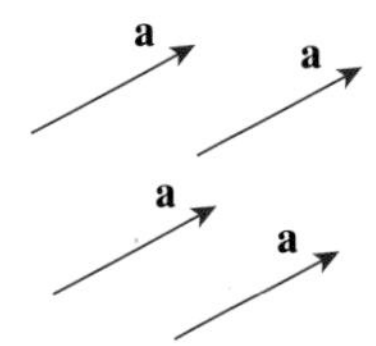

Fig. 6.10
Equal vectors

The vector $\mathbf{a} = \begin{pmatrix} 3 \\ 4 \end{pmatrix}$ given in Cartesian component form can be drawn in any position. It does not have to 'start at the origin': this illustrated in Figure 6.11 below.

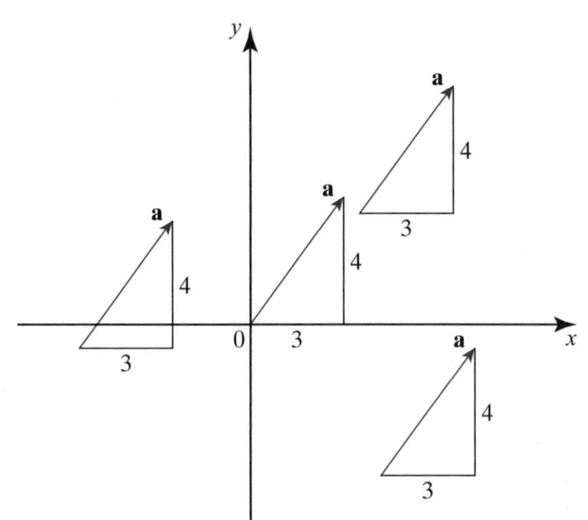

Fig. 6.11
Free vectors

Negative of a vector

The negative of a vector **a** written as −**a** and this has the same magnitude as **a** both points in the opposite direction to **a** as is illustrated in Figure 6.12.

Therefore given the two vectors $\mathbf{b} = \begin{pmatrix} 5 \\ 2 \end{pmatrix}$ and $\mathbf{c} = \begin{pmatrix} -5 \\ -2 \end{pmatrix}$

Then $\begin{pmatrix} -5 \\ -2 \end{pmatrix} = -1 \begin{pmatrix} 5 \\ 2 \end{pmatrix}$ because using the usual algebraic rules, −1 multiplies each of the numbers inside the bracket (column vector) so $\mathbf{c} = -\mathbf{b}$.

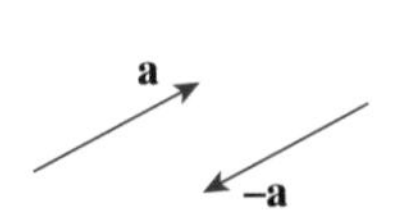

Fig. 6.12
The negative of a vector

Scaling of a vector

Consider the three vectors $\mathbf{b} = \begin{pmatrix} 5 \\ 2 \end{pmatrix}$, $\mathbf{c} = \begin{pmatrix} 10 \\ 4 \end{pmatrix}$ and $\mathbf{d} = \begin{pmatrix} 2.5 \\ 1 \end{pmatrix}$.

You can see that $\mathbf{c} = 2\mathbf{b}$ because $\begin{pmatrix} 10 \\ 4 \end{pmatrix} = 2\begin{pmatrix} 5 \\ 2 \end{pmatrix}$ and $\mathbf{d} = \frac{1}{2}\mathbf{b}$ because $\begin{pmatrix} 2.5 \\ 1 \end{pmatrix} = \frac{1}{2}\begin{pmatrix} 5 \\ 2 \end{pmatrix}$.

The numbers 2 and $\frac{1}{2}$ are scalars (or real numbers).
In each case, the vector **b** is multiplied by a scalar.

Essential notes

Collinear means that the three points *O*, *B* and *C* lie on the same straight line.

If we were to plot the vectors $\mathbf{b} = \overrightarrow{OB} = \begin{pmatrix} 5 \\ 2 \end{pmatrix}$ and $\mathbf{c} = \overrightarrow{OC} = \begin{pmatrix} 10 \\ 4 \end{pmatrix}$ on the same diagram then *O*, *B* and *C* would be **collinear** and *OC* would be 2 *OB*. So **b** and **c** would have the same direction but differ in length.

In general terms this means that when a vector **a** is multiplied by a scalar λ the result λ**a** is a vector

- with magnitude λ times that of **a**
- which has the same direction as **a**.

Exam tips

You should learn this general result and how to apply it.

Example

a) Write each of the following vectors in magnitude-direction form:

$$\mathbf{a}=\begin{pmatrix}6\\8\end{pmatrix}, \mathbf{b}=\begin{pmatrix}3\\1\end{pmatrix}, \mathbf{c}=\begin{pmatrix}12\\16\end{pmatrix}, \mathbf{d}=\begin{pmatrix}-3\\-1\end{pmatrix}, \mathbf{e}=\begin{pmatrix}0.6\\0.8\end{pmatrix}$$

b) What can you say about vectors **a** and **c**?

c) What can you say about vectors **b** and **d**?

d) What can you say about vectors **a** and **e**?

Answer

a) The following table shows the magnitude-direction forms of each vector.

vector	magnitude	direction
$\mathbf{a}=\begin{pmatrix}6\\8\end{pmatrix}$	$\sqrt{6^2+8^2}=\sqrt{100}=10$	$\tan^{-1}\left(\frac{8}{6}\right)=53.13°$
$\mathbf{b}=\begin{pmatrix}3\\1\end{pmatrix}$	$\sqrt{3^2+1^2}=\sqrt{10}$	$\tan^{-1}\left(\frac{1}{3}\right)=18.43°$
$\mathbf{c}=\begin{pmatrix}12\\16\end{pmatrix}$	$\sqrt{12^2+16^2}=\sqrt{400}=20$	$\tan^{-1}\left(\frac{16}{12}\right)=53.13°$
$\mathbf{d}=\begin{pmatrix}-3\\-1\end{pmatrix}$	$\sqrt{(-3)^2+(-1)^2}=\sqrt{10}$	$180°+\tan^{-1}\left(\frac{1}{3}\right)=198.43°$
$\mathbf{e}=\begin{pmatrix}0.6\\0.8\end{pmatrix}$	$\sqrt{0.6^2+0.8^2}=\sqrt{1}=1$	$\tan^{-1}\left(\frac{0.8}{0.6}\right)=53.13°$

b) $\mathbf{a}=\begin{pmatrix}6\\8\end{pmatrix}$ and $\mathbf{c}=\begin{pmatrix}12\\16\end{pmatrix}=2\begin{pmatrix}6\\8\end{pmatrix}$ so $\mathbf{c}=2\mathbf{a}$ which means that **c** is in the same direction as **a** and the magnitude of **c** is twice the magnitude of **a**, or **a** is $\frac{1}{2}$ the magnitude of **c**:

$$\mathbf{a}=\frac{1}{2}\mathbf{c}$$

Method notes

To find the magnitude of the vector use Pythagoras. To find the direction of the vector use trigonometry.

Continued on the next page

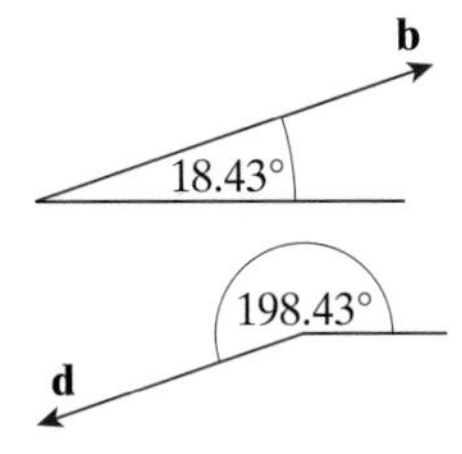

Fig. 6.13
Vectors in the opposite direction

c) $\mathbf{b} = \begin{pmatrix} 3 \\ 1 \end{pmatrix}$ and $\mathbf{d} = \begin{pmatrix} -3 \\ -1 \end{pmatrix}$ so $\mathbf{b} = -\mathbf{d}$ therefore it is in the opposite direction to **d** and the magnitude of **d** is equal to the magnitude of **b** $\Rightarrow \mathbf{b} = -\mathbf{d}$ as shown in Figure 6.13

d) $\mathbf{a} = \begin{pmatrix} 6 \\ 8 \end{pmatrix}$ and $\mathbf{e} = \begin{pmatrix} 0.6 \\ 0.8 \end{pmatrix}$ so $\begin{pmatrix} 6 \\ 8 \end{pmatrix} = 10 \begin{pmatrix} 0.6 \\ 0.8 \end{pmatrix}$ therefore $\mathbf{a} = 10\mathbf{e}$ which means the magnitude of **a** is 10 times the magnitude of **e** and is in the same direction as **e**.

Unit vectors

In the last example, by using Pythagoras, the vector $\mathbf{e} = \begin{pmatrix} 0.6 \\ 0.8 \end{pmatrix}$ has magnitude 1

This is called a **unit vector** because it has a magnitude of 1

Method notes

By Pythagoras $(0.6)^2 + (0.8)^2 = 0.64 + 0.36 = 1$ so magnitude of $\mathbf{e} = \sqrt{1} = 1$

Stop and think 1

If $\mathbf{a} = \begin{pmatrix} 6 \\ 8 \end{pmatrix}$ *find a unit vector in the same direction as* **a**.

Example

Given a vector $\mathbf{b} = \begin{pmatrix} 5 \\ 12 \end{pmatrix}$ find a general formula for the unit vector in the direction of **b** in terms of **b**.

Answer

Step 1: Find the length of **b** so $|\mathbf{b}| = \sqrt{5^2 + 12^2} = \sqrt{169} = 13$

Step 2: Let the required unit vector be **u** (which has length 1) so as **b** and **u** are in the same direction then from step 1, $\mathbf{b} = 13\mathbf{u}$.

Step 3: Make **u** the subject of the equation in step 2:

if $\mathbf{b} = 13\mathbf{u}$ then $\mathbf{u} = \frac{1}{13}\mathbf{b}$

Step 4: Substitute the result from step 1 into the equation from step 3:

$$\mathbf{u} = \frac{\mathbf{b}}{|\mathbf{b}|}.$$

This is the general result for a unit vector in the direction of **b**.

Unit vectors are useful for denoting directions in space. In particular the unit vectors pointing in the positive x- and y-directions are given special labels **i** and **j** respectively. These are called the **Cartesian unit vectors**.

Any two dimensional vector can be written in terms of the unit vectors **i** and **j** and using the Cartesian components as explained in the following example.

Example

Write the vector $\mathbf{a} = \begin{pmatrix} 2 \\ 1.5 \end{pmatrix}$ in terms of the unit vectors **i** and **j**.

Answer

Step 1: Draw the vector $\mathbf{a} = \begin{pmatrix} 2 \\ 1.5 \end{pmatrix}$ with an x component $= 2$ and y component $= 1.5$ and the direction of the unit vectors **i** and **j** along the positive x and y axes respectively as shown in Figure 6.14

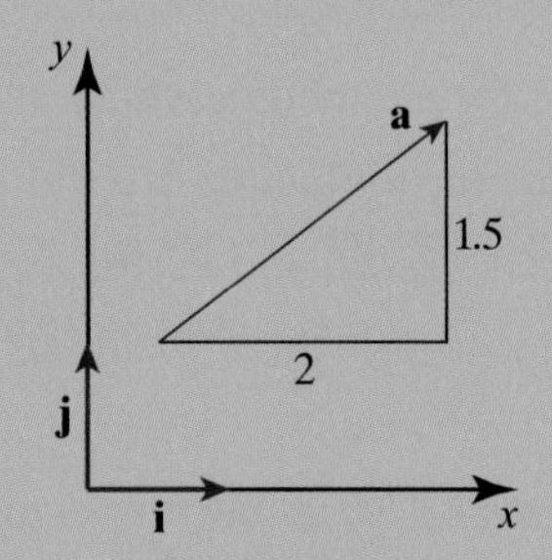

Fig. 6.14
The vector $2\mathbf{i} + 1.5\mathbf{j}$

Step 2: The x component vector is of length 2 and is in the same direction as **i** so using vector scaling it is twice the length of **i** and therefore $= 2\mathbf{i}$.

Step 3: The y component vector is of length 1.5 and is in the same direction as **j** so using vector scaling it is 1.5 times the length of **j** and therefore $= 1.5\mathbf{j}$.

Step 4: Use the results from steps 2 and 3 in the triangle law of vector addition for **a** therefore $\mathbf{a} = 2\mathbf{i} + 1.5\mathbf{j}$.

Vector operations

Adding vectors geometrically

You have seen from the 'mathematical journey' example on page 90 that two displacement vectors can be added according to the triangle law of addition. This rule applies for all vector quantities which may not at first seem to have a 'common' point.

As explained previously in this chapter, vectors are sometimes called 'free' vectors as they are not fixed in space. This means we can move them in our diagrams, without changing their direction and magnitude, in a way which will allow us to use the triangle law of addition for problem solving.

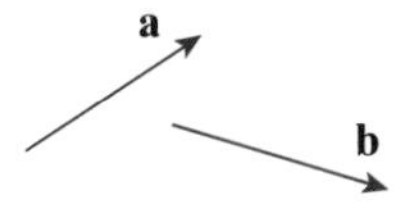

Fig. 6.15
Two free vectors **a** and **b**.

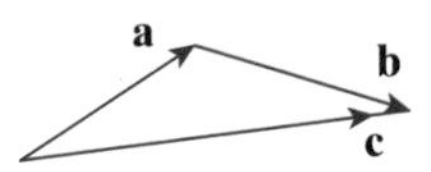

Fig. 6.16
The triangle law of vector addition.

Two vectors are added geometrically according to the **triangle law of vector addition.**

Given two vectors **a** and **b** as shown in Figure 6.15 the sum of **a** and **b** is found by moving vector **b** so that its 'tail' meets the 'head' of vector **a** as shown in Figure 6.16.

We can then use the triangle law of addition so if we complete the triangle and label the vector from the tail of **a** to the head of **b**, as **c**, then the vector sum of **a** and **b** is written as $\mathbf{a} + \mathbf{b} = \mathbf{c}$.

The sum of two vectors (or more than two vectors) is called the **resultant** vector so in this case **c** is the **resultant** vector.

Vectors can be used to prove geometric properties as the next example illustrates.

Method notes

When drawing the triangle of vectors you **must** ensure that for the vectors being added, the **head** of the first vector becomes the **tail** of the second vector.

Example

The points A, B and C form a triangle where $\overrightarrow{AB} = \mathbf{a}$ and $\overrightarrow{AC} = \mathbf{b}$.

The points P and Q are the midpoints of AB and AC.

a) Write the vectors $\overrightarrow{AP}$, $\overrightarrow{AQ}$, $\overrightarrow{BC}$ and $\overrightarrow{PQ}$ in terms of **a** and **b**.

b) What can you deduce about the PQ and BC?

Answer

a) **Step 1:** Draw a triangle ABC, show the vectors **a** and **b** and mark the mid-points of AB and AC as P and Q respectively as shown in Figure 6.17:

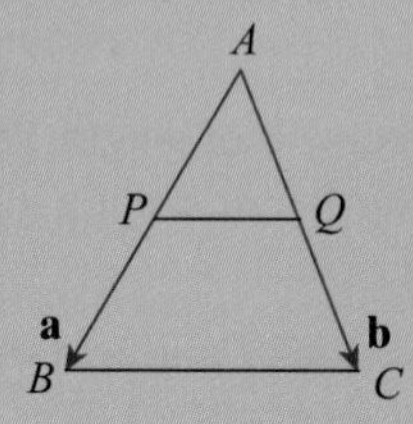

Fig. 6.17
The triangle ABC with midpoints of the sides marked as P and Q.

Step 2: Use scaling of vectors to find $\overrightarrow{AP}$.

Since the magnitude (i.e. length) of $\overrightarrow{AP}$ is half the magnitude of $\overrightarrow{AB}$, and $\overrightarrow{AP}$ is parallel to $\overrightarrow{AB}$ we can write:

$$\overrightarrow{AP} = \frac{1}{2}\overrightarrow{AB} = \frac{1}{2}\mathbf{a}$$

Step 3: Use scaling of vector to find $\overrightarrow{AQ}$ so by a similar argument as used in step 2:

$$\overrightarrow{AQ} = \frac{1}{2}\overrightarrow{AC} = \frac{1}{2}\mathbf{b}$$

Step 4: Use the triangle rule for vector addition in triangle *ABC*:

$$\mathbf{a} + \overrightarrow{BC} = \mathbf{b} \Rightarrow \overrightarrow{BC} = \mathbf{b} - \mathbf{a}$$

Step 5: Use the triangle rule for vector addition in triangle *APQ*:

$$\overrightarrow{AP} + \overrightarrow{PQ} = \overrightarrow{AQ} \Rightarrow \overrightarrow{PQ} = \overrightarrow{AQ} - \overrightarrow{AP}$$

Step 6: Substitute the results from steps 2 and 3 into the equation from step 5:

$$\overrightarrow{PQ} = \frac{1}{2}\mathbf{b} - \frac{1}{2}\mathbf{a} = \frac{1}{2}(\mathbf{b} - \mathbf{a})$$

b) From part a) $\overrightarrow{BC} = \mathbf{b} - \mathbf{a}$ and $\overrightarrow{PQ} = \frac{1}{2}(\mathbf{b} - \mathbf{a})$

$$\Rightarrow \overrightarrow{PQ} = \frac{1}{2}\overrightarrow{BC}$$

therefore the line *PQ* is parallel to the line *BC* and *PQ* is half the magnitude or length of *BC*.

Example

Given that *ABCDE* is a pentagon, use the triangle rule of vector addition to show that $\overrightarrow{AB} + \overrightarrow{BC} + \overrightarrow{CD} + \overrightarrow{ED} + \overrightarrow{AD} = 2\,\overrightarrow{AD}$.

Answer

Step 1: Draw the pentagon *ABCDE* and label the vertices in a clockwise order as shown in Figure 6.18 below.

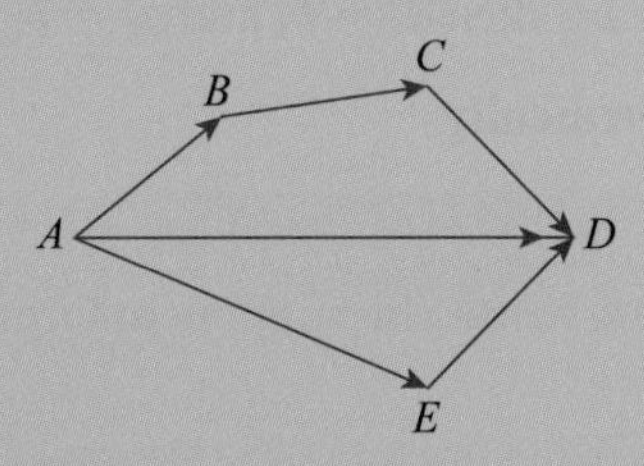

Fig. 6.18
The pentagon *ABCDE*

Step 2: Join *A* to *C* and use the vector law of addition in triangle *ABC* so $\overrightarrow{AB} + \overrightarrow{BC} = \overrightarrow{AC}$.

Step 3: Use the vector law of addition in triangle *ACD* so $\overrightarrow{AC} + \overrightarrow{CD} = \overrightarrow{AD}$.

Step 4: Substitute the result for $\overrightarrow{AC}$ from step 2 into the equation in step 3:

$$\overrightarrow{AB} + \overrightarrow{BC} + \overrightarrow{CD} = \overrightarrow{AD}$$

Step 5: Use the vector law of addition for triangle *AED* so $\overrightarrow{AE} + \overrightarrow{ED} = \overrightarrow{AD}$.

Step 6: Add the results from steps 4 and 5:

$$\overrightarrow{AB} + \overrightarrow{BC} + \overrightarrow{CD} + \overrightarrow{ED} + \overrightarrow{AE} = \overrightarrow{AD} + \overrightarrow{AD} = 2\,\overrightarrow{AD}$$

which is the required result.

Method notes

When labelling the vertices of any shape it is important to do this in either clockwise or anti -clockwise order. When labelling you must not miss out vertices or 'jump' across the shape.

Equal vectors

Example

Prove that if **a** and **b** are non-parallel, non-zero vectors and $\lambda\mathbf{a} + \mu\mathbf{b} = \alpha\mathbf{a} + \beta\mathbf{b}$ then $\lambda = \alpha$ and $\mu = \beta$ where α, β, λ and μ are constants.

Answer

Step 1: Given $\lambda\mathbf{a} + \mu\mathbf{b} = \alpha\mathbf{a} + \beta\mathbf{b}$ simplify algebraically so $\lambda\mathbf{a} - \alpha\mathbf{a} = \beta\mathbf{b} - \mu\mathbf{b}$ therefore by factorising $(\lambda - \alpha)\mathbf{a} = (\beta - \mu)\mathbf{b}$.

Step 2: From the result in step 1 use scaling of vectors so since $(\lambda - \alpha)$ and $(\beta - \mu)$ are constants then **a** and **b** must be in the same direction therefore **a** is parallel to **b**.

Step 3: Use the given fact that **a** and **b** are non-parallel, non-zero vectors and the results from steps 1 and 2 to deduce that for $(\lambda - \alpha)\mathbf{a} = (\beta - \mu)\mathbf{b}$ to hold true this can only happen if:

$\lambda - \alpha = 0$ and $\beta - \mu = 0$ therefore $\lambda = \alpha$ and $\mu = \beta$.

The example leads to the general result that the addition of two vectors $\lambda\mathbf{a} + \mu\mathbf{b}$ is equal to the addition of two other vectors $\alpha\mathbf{a} + \beta\mathbf{b}$ so

$\lambda\mathbf{a} + \mu\mathbf{b} = \alpha\mathbf{a} + \beta\mathbf{b}$ only if $\lambda = \alpha$ and $\mu = \beta$.

Exam tips

You should learn this general result as its application in problem solving may be tested in the examination.

An important special case of this result is that:

if $a_1\mathbf{i} + a_2\mathbf{j} = b_1\mathbf{i} + b_2\mathbf{j}$ then $a_1 = b_1$ and $a_2 = b_2$

Adding vectors algebraically

Example

Two vectors **a** and **b** are written in component form as $\mathbf{a} = 2\mathbf{i} + 3\mathbf{j}$ and $\mathbf{b} = \mathbf{i} - \mathbf{j}$.

a) Draw a diagram showing the vectors **a**, **b** and **a** + **b**.

b) Write down the component form of **a** + **b**.

Answer

a) **Step 1:** Draw the vector $\mathbf{a} = 2\mathbf{i} + 3\mathbf{j}$ as the resultant of two component vectors, one is 2 units in the positive x direction and the other is 3 units in the positive y direction as shown in Figure 6.19 below.

Step 2: Draw the vector $\mathbf{b} = \mathbf{i} - \mathbf{j}$ as the resultant of two component vectors, one is 1 unit in the positive x direction and the other is 1 unit in the positive y direction as shown in Figure 6.19 below.

Step 3: Use the vector law of addition to complete the triangle with sides representing **a** and **b**, to give the vector **a** + **b** with the tail of **b** is joined to the head of **a**, as shown in Figure 6.19 opposite.

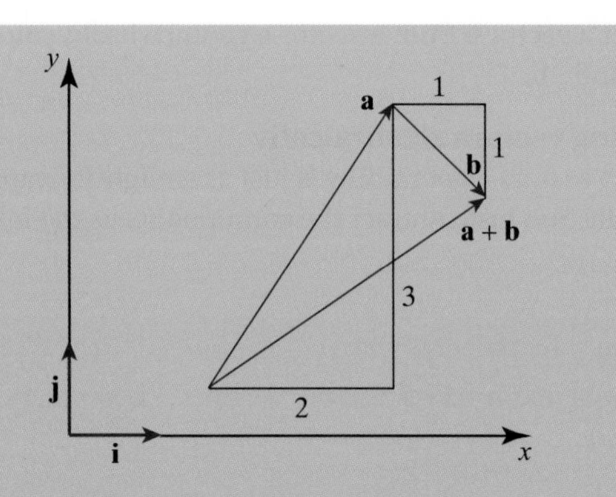

Fig. 6.19
The vector $\mathbf{a} + \mathbf{b}$.

b) **Step 1:** From the diagram write down the component form of **a**

$\mathbf{a} = 2\mathbf{i} + 3\mathbf{j}$ and $\mathbf{b} = \mathbf{i} - \mathbf{j}$ so $\mathbf{a} + \mathbf{b} = 2\mathbf{i} + 3\mathbf{j} + \mathbf{i} - \mathbf{j}$

Step 2: Simplify the answer in step 1 algebraically so

$\mathbf{a} + \mathbf{b} = (2 + 1)\mathbf{i} + (3 - 1)\mathbf{j}$

therefore $\mathbf{a} + \mathbf{b} = 3\mathbf{i} + 2\mathbf{j}$

From the example above, you can see that to find the sum of the vectors **a** and **b** given in component form we just need to add the corresponding components of **a** to the components of **b**.

Exam tips

You should learn this general result as its application in problem solving may be tested in the examination.

In general, two vectors $\mathbf{a} = \begin{pmatrix} a_1 \\ a_2 \end{pmatrix} = a_1\mathbf{i} + a_2\mathbf{j}$ and $\mathbf{b} = \begin{pmatrix} b_1 \\ b_2 \end{pmatrix} = b_1\mathbf{i} + b_2\mathbf{j}$ are added algebraically by adding the components:

$$\mathbf{a} + \mathbf{b} = \begin{pmatrix} a_1 + b_1 \\ a_2 + b_2 \end{pmatrix} = (a_1 + \mathrm{b}_1)\mathbf{i} + (a_2 + \mathrm{b}_2)\mathbf{j}$$

Subtracting vectors geometrically

Method notes

The negative of a vector was explained earlier in the chapter. Vectors **b** and **–b** are parallel vectors , equal magnitude but opposite in direction.

Example

Given the vectors **a** and **b** show on a diagram the vector $\mathbf{a} - \mathbf{b}$.

Answer

Step 1: Rewrite $\mathbf{a} - \mathbf{b}$ algebraically so $\mathbf{a} - \mathbf{b} = \mathbf{a} + (-\mathbf{b})$

Step 2: Use the vector law of addition to complete the triangle with sides representing **a** and **–b** to give the vector $\mathbf{a} - \mathbf{b}$ with the tail of the vector (**–b**) joined to the head of the vector of **a** as shown in Figure 6.20 below.

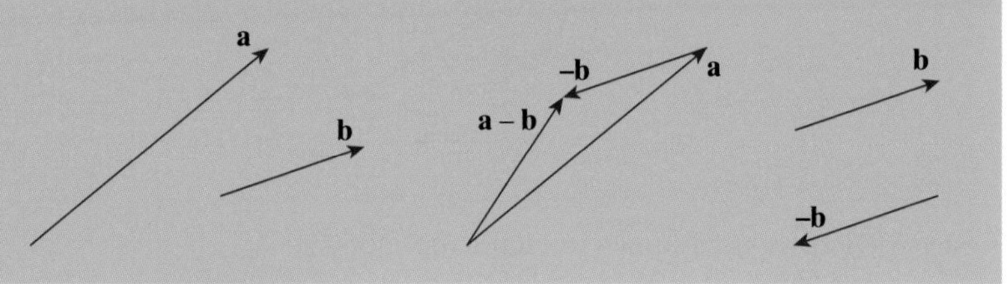

Fig. 6.20
Subtracting two vectors **a** and **b**.

Continued on the next page

To subtract a vector **b** from a vector **a** geometrically, you just add the vectors **a** and −**b**.

Subtracting vectors algebraically

To subtract vectors algebraically is just as straightforward as adding vectors algebraically. You just subtract the components as the following example illustrates.

Example

If $\mathbf{a} = 2\mathbf{i} + 3\mathbf{j}$ and $\mathbf{b} = \mathbf{i} - \mathbf{j}$ find the vector $\mathbf{a} - \mathbf{b}$ in component form.

Answer

Use the general result $\mathbf{a} - \mathbf{b} = \mathbf{a} + (-\mathbf{b}) = \begin{pmatrix} a_1 - b_1 \\ a_2 - b_2 \end{pmatrix} = (a_1 - b_1)\mathbf{i} + (a_2 - b_2)\mathbf{j}$

where $\mathbf{a} = a_1\mathbf{i} + a_2\mathbf{j}$ and $\mathbf{b} = b_1\mathbf{i} + b_2\mathbf{j}$ so

if $\mathbf{a} = 2\mathbf{i} + 3\mathbf{j}$ and $\mathbf{b} = \mathbf{i} - \mathbf{j}$ then $a_1 = 2$ and $a_2 = 3$, $b_1 = 1$ and $b_2 = -1$

therefore $\mathbf{a} - \mathbf{b} = (2 - 1)\mathbf{i} + (3 - (-1))\mathbf{j} = \mathbf{i} + 4\mathbf{j}$

Method notes

You must learn the two methods for subtracting vectors i.e. geometrically and algebraically.

Position vectors

The position vector of a point P is the vector drawn from the origin to the point P. It is a special vector which is not a free vector and always starts from the origin. Such a vector is also often called a **bound vector**. It is usual to call this position vector of the point P, **p**. Similarly it is usual to call the position vector of the point Q from the origin O, **q**.

The letter **r** is sometimes reserved to denote a position vector for a point with Cartesian coordinates (x, y) but any letter can be used. So if a point has x- and y-coordinates $(4, -1)$ then in component form, the position vector of this point (relative to the origin O of Cartesian coordinates) is $\mathbf{r} = 4\mathbf{i} - \mathbf{j}$

Example

Given two points P and Q with position vectors **p** and **q**, show on a diagram the vector $\mathbf{q} - \mathbf{p}$.

Answer

As this is a question involving position vectors then the origin O must be the starting point for each of the two vectors **p** and **q**.

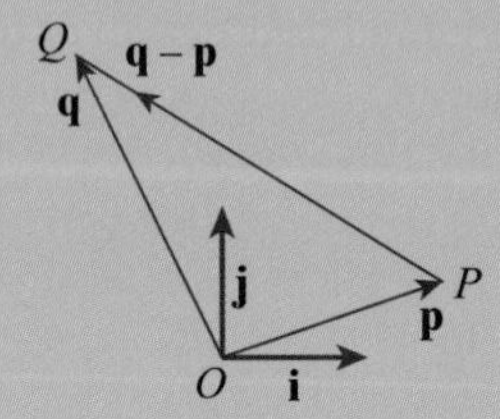

Fig. 6.21
The vector $\mathbf{q} - \mathbf{p}$.

So $\mathbf{p} = \overline{OP}$ $\mathbf{q} = \overrightarrow{OQ}$ as shown in Figure 6.21 above.

Step 1: Use the negative of a vector (if $\mathbf{p} = \overrightarrow{OP}$ then $-\mathbf{p} = \overrightarrow{PO}$) in the triangle law of addition:

$\overrightarrow{PQ} = \overrightarrow{PO} + \overrightarrow{OQ}$

Therefore $\overrightarrow{PQ} = -\mathbf{p} + \mathbf{q} = \mathbf{q} - \mathbf{p}$

Step 2: This is illustrated by the triangle *OPQ* in Figure 6.21.

The displacement vector $\overrightarrow{PQ}$ gives the position of *Q* relative to *P*.

Example

Two points *P* and *Q* have position vectors $\mathbf{p} = -2\mathbf{i} + 3\mathbf{j}$ and $\mathbf{q} = 5\mathbf{i} + 8\mathbf{j}$ respectively.

a) Find the distance between *P* and *Q* and the direction $\overrightarrow{PQ}$ as a bearing of the position of *Q* relative to *P*.

b) Write down a unit vector in the direction of $\overrightarrow{PQ}$.

Answer

a) **Step 1:** Find $\overrightarrow{PQ}$ which is $\mathbf{q} - \mathbf{p}$ (and is the position of *Q* relative to *P*) using component form so

$\mathbf{q} - \mathbf{p} = (5\mathbf{i} + 8\mathbf{j}) - (-2\mathbf{i} + 3\mathbf{j}) = 7\mathbf{i} + 5\mathbf{j}$

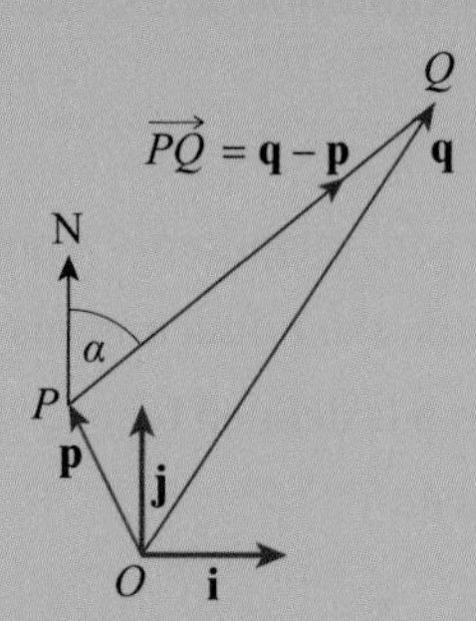

Fig. 6.22
The bearing of *Q* relative to *P*.

Step 2: Find the distance between *P* and *Q* which is the magnitude of the vector $\mathbf{q} - \mathbf{p}$ so

$|\mathbf{q} - \mathbf{p}| = \sqrt{7^2 + 5^2} = \sqrt{74} = 8.6$

Step 3: Find the direction of $\overrightarrow{PQ}$ which is the angle $(90 - \alpha)^\circ$ using the components of $\overrightarrow{PQ}$

so $\tan (90 - \alpha)^\circ = \dfrac{5}{7} \Rightarrow (90 - \alpha)^\circ = 35.5^\circ$ therefore $\alpha = 54.5^\circ$

Therefore the bearing of *Q* relative to *P* (which is always measured from North clockwise) is the angle α shown in Figure 6.22 and is 054.5°.

Method notes

The direction of a vector given as $\begin{pmatrix} 6 \\ 8 \end{pmatrix}$ is $\tan^{-1}\left(\dfrac{8}{6}\right)$ and is the angle made with the positive direction of the *x*-axis. This was discussed earlier in the chapter. In this example the direction angle is $(90 - \alpha)^\circ$.

Continued on the next page

b) A unit vector in the direction of $\overrightarrow{PQ}$ is given by $\dfrac{\overrightarrow{PQ}}{|\overrightarrow{PQ}|}$ as

explained earlier in this chapter.

So if $\overrightarrow{PQ} = 7\mathbf{i} + 5\mathbf{j}$ then $|\overrightarrow{PQ}| = \sqrt{74}$

therefore $\dfrac{\overrightarrow{PQ}}{|\overrightarrow{PQ}|} = \dfrac{7\mathbf{i} + 5\mathbf{j}}{\sqrt{74}} = \dfrac{7}{\sqrt{74}}\mathbf{i} + \dfrac{5}{\sqrt{74}}\mathbf{j}$ which is the unit

vector in the direction of $|\overrightarrow{PQ}|$.

Products of vectors

Having defined the addition and subtraction of two vectors and the multiplication of a vector by a scalar (called scaling) we now look at how to multiply two vectors. We can never divide one vector by another because it is not defined and has no meaning.

There are two ways of multiplying two vectors. One leads to a quantity which is a scalar which is called the **scalar product**. The other leads to a quantity which is a vector and is called the **vector product**. These two products are defined because of their applications to physical quantities such as displacement and force.

In Core 4 you only need to know how to use the scalar product.

The scalar product

Definition

Given two vectors **a** and **b** with magnitudes a and b respectively

and with the angle θ between their directions as shown in Figure 6.23,

the scalar product of **a** and **b** is defined by

$\mathbf{a} \cdot \mathbf{b} = |\mathbf{a}|\,|\mathbf{b}| \cos\theta = ab\cos\theta$

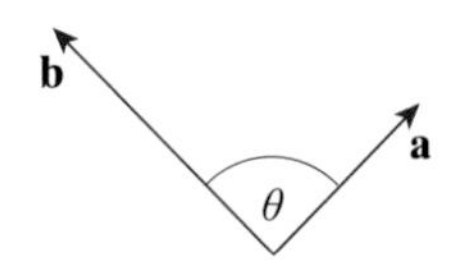

Fig. 6.23
The vectors **a** and **b** from a common point.

This notation is very important and the dot between the two vectors must be clearly shown; the result is a scalar quantity.

When reading the formula for the scalar product, it is pronounced 'a dot b'.

Essential notes

To use the scalar product ,the two vectors **must** be drawn so that their directions are outwards from the common point ie their two tails meet at the common point.

θ is then the angle between the two vectors.

Example

Three vectors **a, b** and **c** are shown in the diagram.

Find the values of

a) $\mathbf{a} \cdot \mathbf{b}$

b) $\mathbf{b} \cdot \mathbf{c}$

c) $\mathbf{a} \cdot \mathbf{c}$

Answer

a) **Step 1:** This is a question using the dot product therefore check that in the diagram representing the vectors (Figure 6.24) the 'tails' of the vectors meet at the common point which they do!

Step 2: From the diagram write down the magnitude of the three vectors

So $|\mathbf{a}| = 3$, $|\mathbf{b}| = 7$ and $|\mathbf{c}| = 4$

Step 3: Use the definition of $\mathbf{a} \cdot \mathbf{b} = |\mathbf{a}|\,|\mathbf{b}| \cos\theta$ with $\theta = 55^{\circ}$ and the result from step 2 so

$\mathbf{a} \cdot \mathbf{b} = 3 \times 7 \times \cos 55^{\circ} = 21 \cos 55^{\circ} = 12.05$

b) Use the definition of $\mathbf{b} \cdot \mathbf{c} = |\mathbf{b}|\,|\mathbf{c}| \cos\theta$ with $\theta = 35^{\circ}$ and the magnitudes of the vectors

so $\mathbf{b} \cdot \mathbf{c} = |\mathbf{b}|\,|\mathbf{c}| \cos^{\circ} = 7 \times 4 \times \cos 35^{\circ} = 28 \cos 35^{\circ} = 22.94$

Use the definition of $\mathbf{a} \cdot \mathbf{c} = |\mathbf{a}|\,|\mathbf{c}| \cos\theta$ with $\theta = 90^{\circ}$ and the magnitudes of the vectors so $\mathbf{a} \cdot \mathbf{c} = |\mathbf{a}|\,|\mathbf{c}| \cos\theta = 3 \times 4 \times \cos 90^{\circ} = 12 \cos 90^{\circ} = 0$

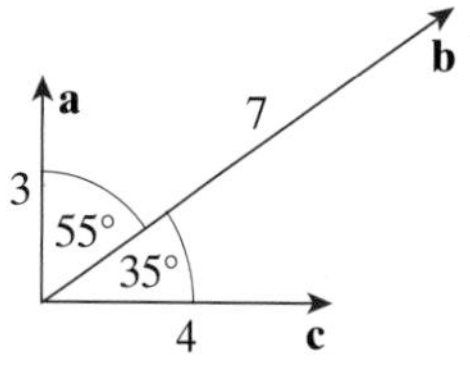

Fig. 6.23
The vectors **a**, **b** and **c** from a common point.

Perpendicular vectors

If the angle between two vectors **a** and **b** is 90° they are **perpendicular vectors**.

The scalar product of **a** and **b** is then $\mathbf{a} \cdot \mathbf{b} = |\mathbf{a}|\,|\mathbf{b}| \cos 90^{\circ}$ and since $\cos 90^{\circ} = 0$ it follows that $\mathbf{a} \cdot \mathbf{b} = 0$

Conversely, if $\mathbf{a} \cdot \mathbf{b} = 0$ then either $\mathbf{a} = \mathbf{0}$, or $\mathbf{b} = \mathbf{0}$ or **a** is perpendicular to **b**.

It is unlikely that $\mathbf{a} = \mathbf{0}$ or $\mathbf{b} = \mathbf{0}$ (of magnitude 0) so you can usually assume the following as a test for perpendicular vectors:

if $\mathbf{a} \cdot \mathbf{b} = 0$ then **a** is perpendicular to **b**.

Exam tips

You should learn how to apply the dot product result for perpendicular vectors as it may be tested in the examination.

Scalar products and Cartesian components

The property that two **perpendicular vectors** have zero scalar product is very useful when working with vectors given in Cartesian component form.

Since **i** and **j** were defined earlier in this chapter as unit vectors in the directions of the x and y axis respectively then they are perpendicular vectors so we can deduce from the scalar product definition that

$\mathbf{i} \cdot \mathbf{j} = 1 \times 1 \times \cos 90° = 0$

$\mathbf{j} \cdot \mathbf{i} = 1 \times 1 \times \cos 90° = 0$

Example

Given the two vectors $\mathbf{a} = a_1\mathbf{i} + a_2\mathbf{j}$ and $\mathbf{b} = b_1\mathbf{i} + b_2\mathbf{j}$ evaluate $\mathbf{a} \cdot \mathbf{b}$.

Answer

Step 1: Substitute for **a** and **b** in terms of **i** and **j** into $\mathbf{a} \cdot \mathbf{b}$ so

$$\mathbf{a} \cdot \mathbf{b} = (a_1\mathbf{i} + a_2\mathbf{j}) \cdot (b_1\mathbf{i} + b_2\mathbf{j})$$

Step 2: Apply the usual algebraic method of eliminating brackets to the equation in step 1 so

$$\mathbf{a} \cdot \mathbf{b} = (a_1\mathbf{i} + a_2\mathbf{j}) \cdot (b_1\mathbf{i} + b_2\mathbf{j}) = a_1b_1\mathbf{i} \cdot \mathbf{i} + a_1b_2\mathbf{i} \cdot \mathbf{j} + a_2b_1\mathbf{j} \cdot \mathbf{i} + a_2b_2\mathbf{j} \cdot \mathbf{j}$$

Step 3: Because **i** and **j** are perpendicular vectors, $\mathbf{i} \cdot \mathbf{j} = 0$ and $\mathbf{j} \cdot \mathbf{i} = 0$ so substitute these results into the equation in step 2 therefore

$$\mathbf{a} \cdot \mathbf{b} = a_1b_1\mathbf{i} \cdot \mathbf{i} + 0 + 0 + a_2b_2\mathbf{j} \cdot \mathbf{j}$$

Step 4: Because $\mathbf{i} \cdot \mathbf{i} = \mathbf{j} \cdot \mathbf{j} = 1$ (as shown earlier in the stop and think question 3) substitute these results into the equation in step 3 so

$$\mathbf{a} \cdot \mathbf{b} = a_1b_1 + a_2b_2$$

Exam tips

You should learn how to apply this result as it may be tested in the examination.

Therefore the scalar product of two vectors is equal to the sum of the products of the **i** and **j** components of the two vectors.

This is a very useful result in solving geometric problems involving vectors as the following example illustrates.

Example

Vectors **a, b** and **c** are given by $4\mathbf{i} + 3\mathbf{j}$, $12\mathbf{i} + 5\mathbf{j}$ and $-6\mathbf{i} + 8\mathbf{j}$ respectively.

Find the angle between each pair of vectors.

Answer

Step 1: Each vector is in component form so if $\mathbf{a} = 4\mathbf{i} + 3\mathbf{j}$ then $a_1 = 4$ and $a_2 = 3$ and $\mathbf{b} = 12\mathbf{i} + 5\mathbf{j}$ then $b_1 = 12$ and $b_2 = 5$ and $\mathbf{c} = -6\mathbf{i} + 8\mathbf{j}$ then $c_1 = -6$ and $c_2 = 8$

Step 2: To find the angle θ between **a** and **b** use $\mathbf{a} \cdot \mathbf{b} = a_1b_1 + a_2b_2$ and substitute into this the answers from step 1:

$$\mathbf{a} \cdot \mathbf{b} = 4 \times 12 + 3 \times 5 = 63$$

Step 3: Find $|\mathbf{a}| = \sqrt{16 + 9} = 5$ and $|\mathbf{b}| = \sqrt{144 + 25} = 13$

Step 4: Use the definition $\mathbf{a} \cdot \mathbf{b} = |\mathbf{a}|\,|\mathbf{b}| \cos\theta$, rewrite and then substitute the results from steps 1, 2 and 3:

$$\cos\theta = \frac{\mathbf{a} \cdot \mathbf{b}}{ab} = \frac{63}{5 \times 13} = \frac{63}{65}$$

therefore the angle θ between the vectors **a** and **b** is 14.25°

Step 5: To find the angle α between **b** and **c** use $\mathbf{b} \cdot \mathbf{c} = b_1c_1 + b_2c_2$ and substitute into this the answers from step 1:

$\mathbf{b} \cdot \mathbf{c} = 12 \times (-6) + 5 \times 8 = -32$

Step 6: Find $|\mathbf{c}| = \sqrt{36 + 64} = 10$

Step 7: Use the definition $\mathbf{b} \cdot \mathbf{c} = |\mathbf{b}|\,|\mathbf{c}| \cos\alpha$, rewrite and then substitute the results from steps 3, 5 and 6:

$$\cos\alpha = \frac{\mathbf{b} \cdot \mathbf{c}}{bc} = \frac{-32}{13 \times 10} = \frac{-32}{130}$$

Therefore the angle α between the vectors **b** and **c** is 104.25°

Step 8: To find the angle between **a** and **c** use $\mathbf{a} \cdot \mathbf{c} = a_1c_1 + a_2c_2$ and substitute into this the answers from step1:

$\mathbf{a} \cdot \mathbf{c} = 4 \times (-6) + 3 \times 8 = 0$

Since the scalar product is zero the vectors **a** and **c** are perpendicular therefore the angle between the vectors **a** and **c** is 90°.

Method notes

A negative cosine means the angle is obtuse.

Method notes

Clearly $|\mathbf{a}| \neq 0$ and $|\mathbf{b}| \neq 0$

Example

Two vectors **p** and **q** are given by $\mathbf{p} = 2\mathbf{i} + 3\mathbf{j}$ and $\mathbf{q} = m\mathbf{i} + \mathbf{j}$

Find the value of m so that

a) vectors **p** and **q** are parallel

b) vectors **p** and **q** are perpendicular

c) the angle between **p** and **q** is 60°

Answer

a) **Step 1:** Use the result that two vectors **p** and **q** are parallel if $\mathbf{q} = \lambda\mathbf{p}$ and write the vectors in Cartesian component form:

$m\mathbf{i} + \mathbf{j} = \lambda(2\mathbf{i} + 3\mathbf{j}) = 2\lambda\mathbf{i} + 3\lambda\mathbf{j}$

b) **Step 2:** Use the result for equal vectors, if '$a_1\mathbf{i} + a_2\mathbf{j} = b_1\mathbf{i} + b_2\mathbf{j}$ then $a_1 = b_2$ and $a_2 = b_2$',

in the equation in step 1 so $m = 2\lambda$ and $1 = 3\lambda$

$$\Rightarrow \lambda = \frac{1}{3} \text{ and } m = 2\lambda = \frac{2}{3}$$

Continued on the next page

b) **Step 1:** Given $\mathbf{q} = m\mathbf{i} + \mathbf{j}$ and $\mathbf{p} = 2\mathbf{i} + 3\mathbf{j}$ therefore

$p_1 = 2$, $p_2 = 3$, $q_1 = m$ and $q_2 = 1$

Step 2: Use the result that two vectors **p** and **q** are perpendicular if $\mathbf{p} \cdot \mathbf{q} = 0$ and the general result that $\mathbf{p} \cdot \mathbf{q} = p_1 q_1 + p_2 q_2$ with the result from step 1

so $2m + 3 = 0$

$\Rightarrow m = -\frac{3}{2}$

c) **Step 1:** To find the angle θ between the two vectors **p** and **q**

use $\mathbf{p} \cdot \mathbf{q} = |\mathbf{p}|\,|\mathbf{q}| \cos\theta$ and $\mathbf{p} \cdot \mathbf{q} = p_1 q_1 + p_2 q_2$ where

$p_1 = 2$, $p_2 = 3$, $q_1 = m$ and $q_2 = 1$ so

$$2m + 3 = \sqrt{13} \times \sqrt{m^2 + 1} \times \cos 60 = \frac{\sqrt{13} \times \sqrt{m^2 + 1}}{2}$$

Step 2: Square both sides of the equation in step 1 and solve for m

So $(2m + 3)^2 = \frac{13(m^2 + 1)}{4}$

$\Rightarrow 16m^2 + 48m + 36 = 13m^2 + 13$

$\Rightarrow 3m^2 + 48m + 23 = 0$

$$\Rightarrow m = \frac{48 \pm \sqrt{48^2 - 4 \times 3 \times 23}}{6} = \frac{48 \pm \sqrt{2028}}{6}$$

$\Rightarrow m = -0.49$ or $m = -15.51$ (to 2 d.p.)

Method notes

$\cos 60° = \frac{1}{2}$

Stop and think 2

Given $\mathbf{a} = 3\mathbf{i} + 4\mathbf{j}$ *and* $\mathbf{b} = 9\mathbf{i} + 12\mathbf{j}$:

a) find the angle between **a** *and* **b**.

b) evaluate $\mathbf{i} \cdot \mathbf{i}$ *and* $\mathbf{j} \cdot \mathbf{j}$.

Geometry and vectors in two dimensions

So far in this chapter we have introduced the geometric and algebraic properties of vectors. Some of the examples have also shown how vectors can provide a useful approach to solving problems in geometry.

The rest of the chapter is about the use of vectors when exploring problems associated with straight lines in two and three dimensions.

Example

The position vectors of a set of points are given by the vector equation $\mathbf{r} = \mathbf{a} + t\mathbf{d}$ where $\mathbf{a} = 2\mathbf{i} - \mathbf{j}$, $\mathbf{d} = 3\mathbf{i} + \mathbf{j}$ and t is a scalar **parameter** which may take any value.

a) Write down the position vectors of the points with values of t equal to -2, -1, 0, 1 and 2

b) By drawing a graph, show that the points lie on a straight line.

Answer

a) Substitute the given values of t:

$t = -2 \Rightarrow \mathbf{r} = \mathbf{a} - 2\mathbf{d} = (2\mathbf{i} - \mathbf{j}) - 2(3\mathbf{i} + \mathbf{j}) = -4\mathbf{i} - 3\mathbf{j}$

$t = -1 \Rightarrow \mathbf{r} = \mathbf{a} - \mathbf{d} = (2\mathbf{i} - \mathbf{j}) - (3\mathbf{i} + \mathbf{j}) = -\mathbf{i} - 2\mathbf{j}$

$t = 0 \Rightarrow \mathbf{r} = \mathbf{a} = 2\mathbf{i} - \mathbf{j}$

$t = 1 \Rightarrow \mathbf{r} = \mathbf{a} + \mathbf{d} = (2\mathbf{i} - \mathbf{j}) + (3\mathbf{i} + \mathbf{j}) = 5\mathbf{i}$

$t = 2 \Rightarrow \mathbf{r} = \mathbf{a} + 2\mathbf{d} = (2\mathbf{i} - \mathbf{j}) + 2(3\mathbf{i} + \mathbf{j}) = 8\mathbf{i} + \mathbf{j}$

b) Plot the points given by the position vectors as shown in Figure 6.25 below.

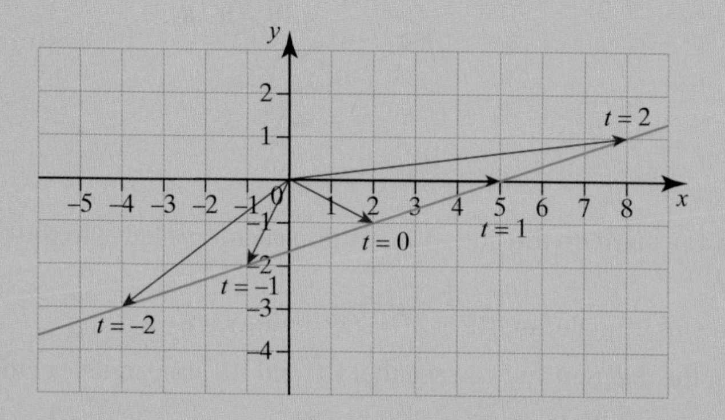

Fig. 6.25
Points plotted from position vectors

We can see that these five points lie on a straight line. As the value of the parameter t changes this gives the series of points, all of which lie on the line. If we were to take more t values this would give more points on the line.

Essential notes

Position vectors were explained earlier in this chapter. In this question the position of any point is given relative to O the origin of Cartesian coordinates. If $\mathbf{r} = -4\mathbf{i} - 3\mathbf{j}$ then this is the position vector of the point $(-4, -3)$.

We can therefore conclude that the equation $\mathbf{r} = \mathbf{a} + t\mathbf{d}$ is an example of the equation of a straight line given in vector terms so we call this the **vector equation of the line**.

Consider the vector between the points with $t = 0$ and $t = 1$, which lies along the line. At $t = 0$ $\mathbf{r} = 2\mathbf{i} - \mathbf{j}$ so let this be the position vector of the point P therefore P is $(2, -1)$ and $\overrightarrow{OP} = 2\mathbf{i} - \mathbf{j}$

At $t = 1$, $\mathbf{r} = 5\mathbf{i}$, so let this be the position vector of the point Q therefore Q is $(5, 0)$ and $\overrightarrow{OQ} = 5\mathbf{i}$.

The vector between the points P and Q is therefore $\overrightarrow{PQ}$ and by the vector triangle law

$\overrightarrow{PQ} = \overrightarrow{PO} + \overrightarrow{OQ} = 5\mathbf{i} - (2\mathbf{i} - \mathbf{j}) = 3\mathbf{i} + \mathbf{j}$.

But $\mathbf{d} = 3\mathbf{i} + \mathbf{j}$ so as P and Q are points on the line then we say the direction vector of the line is $3\mathbf{i} + \mathbf{j}$.

$\mathbf{a} = 2\mathbf{i} - \mathbf{j}$ is the position vector of the point on the line for which $t = 0$

Essential notes

Remember that the vector $\mathbf{d} = 3\mathbf{i} + \mathbf{j}$ gives the direction of the line and the point with position vector $\mathbf{d} = 3\mathbf{i} + \mathbf{j}$ does not lie on the line.

So in this case, the general vector equation $\mathbf{r} = \mathbf{a} + t\mathbf{d}$ is the equation of a straight line through the point with a position vector $\mathbf{a}$ and with a direction vector $\mathbf{d}$.

More generally, suppose that a straight line passes through the two points A and B with position vectors $\mathbf{a}$ and $\mathbf{b}$ respectively relative to an origin O as shown in Figure 6.26.

The line is in the direction of the vector $\overrightarrow{AB}$ and by the vector triangle law

$\overrightarrow{AB} = \overrightarrow{AO} + \overrightarrow{OB} = -\mathbf{a} + \mathbf{b} = \mathbf{b} - \mathbf{a}$

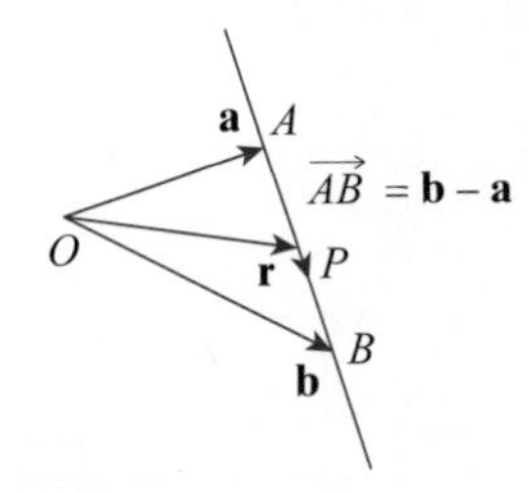

Fig. 6.26
The straight line passing through two points A and B.

Let P be any point on the line with position vector $\mathbf{r}$ relative to the origin O.

By the vector triangle law, $\overrightarrow{AP} = \overrightarrow{AO} + \overrightarrow{OP} = -\mathbf{a} + \mathbf{r} = \mathbf{r} - \mathbf{a}$

But from the diagram you can see that $\overrightarrow{AP}$ and $\overrightarrow{AB}$ are parallel vectors so $\overrightarrow{AP}$ is a scalar multiple of $\overrightarrow{AB}$

$\Rightarrow \mathbf{r} - \mathbf{a} = t(\mathbf{b} - \mathbf{a})$ where t is a scalar parameter.

Simplifying this algebraically gives $\mathbf{r} = (1 - t)\mathbf{a} + t\mathbf{b}$.

$\mathbf{r} = (1 - t)\mathbf{a} + t\mathbf{b}$ is therefore the general parametric equation of a straight line passing through two points with positions vectors $\mathbf{a}$ and $\mathbf{b}$ where t is a scalar parameter and $\mathbf{r}$ is the position vector of a general point on the line.

Essential notes

In problems involving more than one straight line, the letters t and s are often used as parameters or the Greek letters λ and υ. Parameters were explained in Core 3

If we let $\mathbf{d}$ be a vector in the direction of the line passing through the two points A and B then $\mathbf{d} = \mathbf{b} - \mathbf{a}$ since the line is in the direction of the vector $\overrightarrow{AB} = \mathbf{b} - \mathbf{a}$.

From the result $\mathbf{r} - \mathbf{a} = t(\mathbf{b} - \mathbf{a})$ shown earlier and using $\mathbf{d} = \mathbf{b} - \mathbf{a}$ we can rewrite this as $\mathbf{r} = \mathbf{a} + t\mathbf{d}$ where t is a scalar parameter.

$\mathbf{r} = \mathbf{a} + t\mathbf{d}$ is therefore the general parametric equation of a straight line in the direction of a vector $\mathbf{d}$, passing through a point on the line with position vector $\mathbf{a}$, where t is a parameter and $\mathbf{r}$ is the position vector of a general point on the line.

Example

Find the vector equation of the straight line joining the two points *A* and *B* with coordinates (3, 4) and (5, 12) respectively.

Answer

Step 1: Write down the position vectors $\mathbf{a}$ and $\mathbf{b}$ of the points *A* and *B* respectively so as A is (3, 4) then $\mathbf{a} = 3\mathbf{i} + 4\mathbf{j}$ and as *B* is (5, 12) then $\mathbf{b} = 5\mathbf{i} + 12\mathbf{j}$.

Step 2: Find $\overrightarrow{AB}$ using the vector triangle law

so $\overrightarrow{AB} = \mathbf{b} - \mathbf{a} = 5\mathbf{i} + 12\mathbf{j} - (3\mathbf{i} + 4\mathbf{j})$

therefore $\overrightarrow{AB} = 2\mathbf{i} + 8\mathbf{j}$

Step 3: The line is in the direction of $\overrightarrow{AB}$ so let $\mathbf{d} = \overrightarrow{AB}$ and use the result from step 2 so

$\mathbf{d} = 2\mathbf{i} + 8\mathbf{j}$

Step 4: Use $\mathbf{r} = \mathbf{a} + t\mathbf{d}$, the general parametric equation of a line and the results from step 3 so $\mathbf{r} = \mathbf{a} + t(\mathbf{b} - \mathbf{a}) = 3\mathbf{i} + 4\mathbf{j} + t(2\mathbf{i} + 8\mathbf{j})$

This is the vector equation of the line joining the two points *A* and *B*.

Example

Find the vector equation of the straight line through the point *A* with coordinates (−1, 1) and parallel to the vector $2\mathbf{i} + \mathbf{j}$.

Answer

Step 1: Write down the position vector $\mathbf{a}$ of the point $A(-1, 1)$ so $\mathbf{a} = -\mathbf{i} + \mathbf{j}$.

Step 2: Let the direction vector of the line be $\mathbf{d}$ and as this is in the direction of the line which is itself parallel to vector $2\mathbf{i} + \mathbf{j}$ then by scaling $\mathbf{d} = s(2\mathbf{i} + \mathbf{j})$ where s is a scalar parameter.

Step 3: Use $\mathbf{r} = \mathbf{a} + t\mathbf{d}$ the general parametric equation of a line and the results from steps 1 and 2 so

$\mathbf{r} = \mathbf{a} + t\mathbf{d} = -\mathbf{i} + \mathbf{j} + ts(2\mathbf{i} + \mathbf{j})$ which is simplified to $\mathbf{r} = -\mathbf{i} + \mathbf{j} + \lambda(2\mathbf{i} + \mathbf{j})$ where $ts = \lambda$ which is a scalar parameter as t and s are both scalar parameters.

Therefore $\mathbf{r} = -\mathbf{i} + \mathbf{j} + \theta\lambda(2\mathbf{i} + \mathbf{j})$ is the vector equation of the line through the point *A* and parallel to the vector $2\mathbf{i} + \mathbf{j}$.

Essential notes

You can use any symbol to represent a parameter. If t and s are parameters then ts is also a parameter. In this question we use the symbol $\lambda = ts$.

The intersection of two straight lines

When given two straight lines in two dimensions, the lines are either parallel or intersect at one point. The vector equations of the two straight lines provide a useful method for proving whether two straight lines intersect and where the point of intersection occurs.

Example

Show that the two straight lines $\mathbf{r} = \mathbf{i} + 5\mathbf{j} + t(-2\mathbf{i} + \mathbf{j})$ and $\mathbf{r} = 3\mathbf{i} - 4\mathbf{j} + s(-2\mathbf{i} + 5\mathbf{j})$ intersect and find the position vector of the point of intersection.

Answer

Step 1: Let the point of intersection of the two lines be P with position vector $\mathbf{r}$.

If there is a point of intersection the point P must lie on both lines so its position vector must satisfy both vector equations.

Therefore for line 1 given by $\mathbf{r} = \mathbf{i} + 5\mathbf{j} + t(-2\mathbf{i} + \mathbf{j})$ and for line 2 given by $\mathbf{r} = 3\mathbf{i} - 4\mathbf{j} + s(-2\mathbf{i} + 5\mathbf{j})$ equate the position vector $\mathbf{r}$ of the intersection point P so

$\mathbf{i} + 5\mathbf{j} + t(-2\mathbf{i} + \mathbf{j}) = 3\mathbf{i} - 4\mathbf{j} + s(-2\mathbf{i} + 5\mathbf{j})$

$\Rightarrow 1 - 2t = 3 - 2s$

$\Rightarrow 5 + t = -4 + 5s$

Method notes

When two vectors are equal, the **i** components are equal and the **j** components are equal. This was covered earlier in the chapter.

Step 2: Solve the two simultaneous equations in t and s from step 1:

$1 - 2t = 3 - 2s \Rightarrow 1 - 3 + 2s = 2t \Rightarrow -2 + 2s = 2t \Rightarrow -1 + s = t.$

Substitute in the second equation from step 1:

$5 + t = -4 + 5s \Rightarrow 5 + (-1 + s) = -4 + 5s \Rightarrow 8 = 4s \Rightarrow 2 = s.$

As $-1 + s = t$, $t = 1$

Step 3: Substitute the values for T into the line equation:

$\mathbf{r} = \mathbf{i} + 5\mathbf{j} + t(-2\mathbf{i} + \mathbf{j})$ so

$\mathbf{r} = \mathbf{i} + 5\mathbf{j} + t(-2\mathbf{i} + \mathbf{j}) \Rightarrow \mathbf{r} = \mathbf{i} + 5\mathbf{j} + (-2\mathbf{i} + \mathbf{j})$

$\Rightarrow \mathbf{r} = -\mathbf{i} + 6\mathbf{j}$

Therefore the position vector of the point of intersection P is $\mathbf{r} = -\mathbf{i} + 6\mathbf{j}$.

Parallel lines in vector form

In the example above there were unique values for s and t because the two lines did intersect. Suppose that you try to use the same method for two parallel lines for which we know that there is no point of intersection.

For example, consider the lines with vector equations $\mathbf{r} = \mathbf{i} + 2\mathbf{j} + t(-\mathbf{i} + 3\mathbf{j})$ and $\mathbf{r} = 3\mathbf{i} + 4\mathbf{j} + s(2\mathbf{i} - 6\mathbf{j})$

Let the point of intersection be P with position vector $\mathbf{r}$ so equating this position vector on both lines then $\mathbf{i} + 2\mathbf{j} + t(-\mathbf{i} + 3\mathbf{j}) = 3\mathbf{i} + 4\mathbf{j} + s(2\mathbf{i} - 6\mathbf{j})$

Equating the **i** and **j** components

$1 - t = 3 + 2s \Rightarrow t + 2s = -2$

$2 + 3t = 4 - 6s \Rightarrow 3t + 6s = 2 \Rightarrow t + 2s = \frac{2}{3}$

These two equations have no solution because $t + 2s$ cannot equal -2 and $\frac{2}{3}$ simultaneously!

We can deduce that the two straight lines do not have a point of intersection therefore they are parallel lines.

Essential notes

Equal vectors were discussed earlier in this chapter. If vectors are equal then their respective **i** and **j** components are equal.

Another way of deciding whether lines are parallel (and therefore don't intersect) is to compare the direction vector of each line.

In this example line 1 is given by: $\mathbf{r} = \mathbf{i} + 2\mathbf{j} + t(-\mathbf{i} + 3\mathbf{j})$ which means if you compare it with the general form $\mathbf{r} = \mathbf{a} + t\mathbf{d}$ then its direction vector is $\mathbf{d} = (-\mathbf{i} + 3\mathbf{j})$

Line 2 is given by: $\mathbf{r} = 3\mathbf{i} + 4\mathbf{j} + s(2\mathbf{i} - 6\mathbf{j})$ which means if you compare it with the general form $\mathbf{r} = \mathbf{a} + s\mathbf{e}$ its direction vector is $\mathbf{e} = 2\mathbf{i} - 6\mathbf{j} = -2(-\mathbf{i} + 3\mathbf{j})$

So $\mathbf{e} = -2\mathbf{d}$ therefore the vectors $\mathbf{e}$ and $\mathbf{d}$ are parallel vectors because $\mathbf{e}$ is a scaling of $\mathbf{d}$.

Since the direction vectors are parallel the lines must be parallel.

Exam tips

It is worth learning how to interpret geometrically the vector equations of lines. Given $\mathbf{r} = \mathbf{i} + 2\mathbf{j} + t(-\mathbf{i} + 3\mathbf{j})$ this line passes through the point with position vector $\mathbf{i} + 2\mathbf{j}$ so it passes through the point (1, 2) and the line has a direction vector $\mathbf{d} = -\mathbf{i} + 3\mathbf{j}$.

The angle between two straight lines

The vector equations of two straight lines provide a useful method for finding the angle between two straight lines that intersect.

The scalar product gives $\cos\theta = \dfrac{\mathbf{d}\cdot\mathbf{e}}{|\mathbf{d}||\mathbf{e}|}$

where $\mathbf{d}$ and $\mathbf{e}$ are the direction vectors of the two lines and θ is the angle between them.

Essential notes

The scalar product of two vectors was covered earlier in this chapter.

Example

Two straight lines with vector equations

$\mathbf{r} = \mathbf{i} + 5\mathbf{j} + t(-2\mathbf{i} + \mathbf{j})$ and $\mathbf{r} = 3\mathbf{i} - 4\mathbf{j} + s(-2\mathbf{i} + 5\mathbf{j})$ respectively, intersect at a point P.

Find the acute angle between the two lines correct to 1 decimal place.

Answer

Step 1: Write down the direction vector $\mathbf{d}$ of the line

$\mathbf{r} = \mathbf{i} + 5\mathbf{j} + t(-2\mathbf{i} + \mathbf{j})$ so $\mathbf{d} = -2\mathbf{i} + \mathbf{j}$

Step 2: Write down the direction vector $\mathbf{e}$ of the line

$\mathbf{r} = 3\mathbf{i} - 4\mathbf{j} + s(-2\mathbf{i} + 5\mathbf{j})$ so $\mathbf{e} = -2\mathbf{i} + 5\mathbf{j}$

Step 3: Work out $\mathbf{d}\cdot\mathbf{e}$ using the results for $\mathbf{d}$ and $\mathbf{e}$ from steps 1 and 2

so $\mathbf{d}\cdot\mathbf{e} = (-2\mathbf{i} + \mathbf{j})\cdot(-2\mathbf{i} + 5\mathbf{j}) = 4 + 5 = 9$

Step 4: Use the scalar product formula $\cos\theta = \dfrac{\mathbf{d}\cdot\mathbf{e}}{|\mathbf{d}||\mathbf{e}|}$,

the result from step 3 and $|\mathbf{d}| = \sqrt{(-2)^2 + 1^2} = \sqrt{5}$

$|\mathbf{e}| = \sqrt{(-2)^2 + 5^2} = \sqrt{29}$

so $\cos\theta = \dfrac{\mathbf{d}\cdot\mathbf{e}}{|\mathbf{d}||\mathbf{e}|} = \dfrac{9}{\sqrt{5}\sqrt{29}} = 0.7474 \Rightarrow \theta = 41.6°$

Therefore the angle between the two lines is 41.6° correct to 1 decimal place.

Method notes

We know that these two lines intersect at the point (–1, 6) from the previous example. In this example, to find the angle between the lines it is not necessary to find the point of intersection.

Exam tips

When using the scalar product to work out the angle between two lines if $\cos\theta$ is negative remember that θ will be obtuse. Intersecting lines have two angles between them which will always add to give 180°. Read the question carefully and if it asks for the acute angle make sure you give that one!

> **Stop and think 3**
>
> *Find the obtuse angle between the lines given by* $\mathbf{r} = \mathbf{i} + 5\mathbf{j} + t(-2\mathbf{i} + \mathbf{j})$ *and* $\mathbf{r} = 3\mathbf{i} - 4\mathbf{j} + s(-2\mathbf{i} + 5\mathbf{j})$ *respectively,*

Geometry and vectors in three dimensions

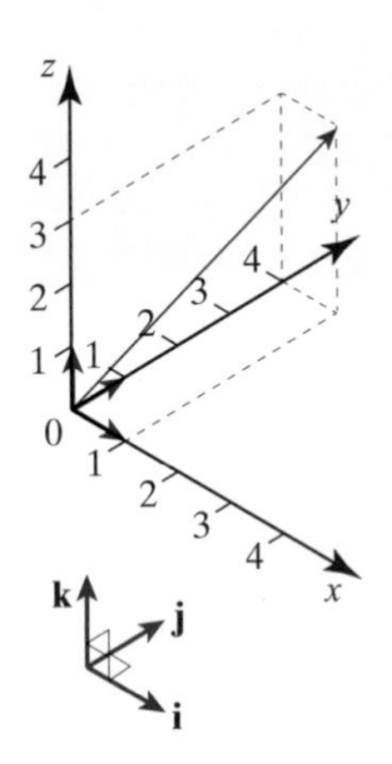

Fig. 6.27
The three-dimensional Cartesian coordinate system.

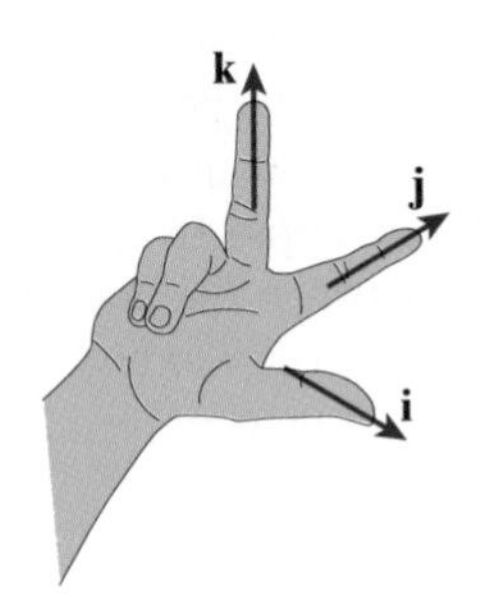

Fig. 6.28
The 'right-hand rule' convention

To describe the position of a point in three dimensions requires a third coordinate axis. Through the origin of a two-dimensional Cartesian coordinate system we draw a third axis which is perpendicular to the x-y plane as shown in the Figure 6.27. This axis is called the z-axis.

It is perpendicular to both the x-axis and the y-axis.

Any point P, such as the point shown in Figure 6.27, can be represented by the perpendicular distances from the x-z, y-z and x-y planes.

The point in Figure 6.27 is written as (1, 4, 3) in this three-dimensional Cartesian coordinate system.

Along this new z-axis we define the unit vector **k**.

The unit vectors **i** and **j** along the x- and y-axes respectively were explained earlier in this chapter.

The coordinate system defined in this way is called a **right-handed coordinate system** as shown in Figure 6.28. If you form (approximate) right angles between your thumb and first two fingers of your right hand, then the unit vector **i** runs along your thumb, the unit vector **j** runs along your first finger and the unit vector **k** runs along your middle finger.

In this newly defined three dimensional system the ideas of vector algebra from two dimensions can be extended as follows:

- any vector **a** can be written in Cartesian component form as $\mathbf{a} = a_1\mathbf{i} + a_2\mathbf{j} + a_3\mathbf{k}$
  (in two dimensions it was $\mathbf{a} = a_1\mathbf{i} + a_2\mathbf{j}$)
- the magnitude of the vector **a** is $\sqrt{a_1^2 + a_2^2 + a_3^2}$
  (in two dimensions it was $|a| = \sqrt{a_1^2 + a_2^2}$)
- the scalar product of two vectors **a** and **b** is: $\mathbf{a} \cdot \mathbf{b} = a_1b_1 + a_2b_2 + a_3b_3$
  (in two dimensions it was: $\mathbf{a} \cdot \mathbf{b} = a_1b_1 + a_2b_2$)
- the equation of a straight line through two points **a** and **b** is $\mathbf{r} = \mathbf{a} + t(\mathbf{b} - \mathbf{a}) = (1 - t)\mathbf{a} + t\mathbf{b}$ where $\mathbf{a} = a_1\mathbf{i} + a_2\mathbf{j} + a_3\mathbf{k}$ and $\mathbf{b} = b_1\mathbf{i} + b_2\mathbf{j} + b_3\mathbf{k}$ and t is a scalar parameter.
  (where in two dimensions $\mathbf{a} = a_1\mathbf{i} + a_2\mathbf{j}$ and $\mathbf{b} = b_1\mathbf{i} + b_2\mathbf{j}$).

  The algebra of vectors can be used to solve problems in three-dimensional geometry as the following examples illustrate.

Example

Three points A, B and C have position vectors $\mathbf{a} = 2\mathbf{i} + \mathbf{j} - \mathbf{k}$, $\mathbf{b} = 3\mathbf{i} + 4\mathbf{j} - 2\mathbf{k}$, $\mathbf{c} = 5\mathbf{i} - \mathbf{j} + 2\mathbf{k}$ respectively, relative to an origin O. The points form a triangle ABC.

a) Write down the vectors $\overrightarrow{AB}$ and $\overrightarrow{AC}$.

b) Calculate the angle between $\overrightarrow{AB}$ and $\overrightarrow{AC}$.

c) Find the area of the triangle ABC.

Answer

a) Using the vector triangle law,

$\overrightarrow{AB} = \mathbf{b} - \mathbf{a} = (3\mathbf{i} + 4\mathbf{j} - 2\mathbf{k}) - (2\mathbf{i} + \mathbf{j} - \mathbf{k}) = \mathbf{i} + 3\mathbf{j} - \mathbf{k}$

$\overrightarrow{AC} = \mathbf{c} - \mathbf{a} = (5\mathbf{i} - \mathbf{j} + 2\mathbf{k}) - (2\mathbf{i} + \mathbf{j} - \mathbf{k}) = 3\mathbf{i} - 2\mathbf{j} + 3\mathbf{k}$

b) **Step 1:** State the scalar product formula for finding the angle between the vectors $\overrightarrow{AB}$ and $\overrightarrow{AC}$ and substitute the results from part a):

$\cos\theta = \dfrac{\overrightarrow{AB} \cdot \overrightarrow{AC}}{|\overrightarrow{AB}||\overrightarrow{AC}|} = \dfrac{(\mathbf{b} - \mathbf{a}) \cdot (\mathbf{c} - \mathbf{a})}{|\mathbf{b} - \mathbf{a}||\mathbf{c} - \mathbf{a}|}$ where θ is the angle between $\overrightarrow{AB}$ and $\overrightarrow{AC}$

Step 2: Find $|\mathbf{b} - \mathbf{a}|$ where $\mathbf{b} - \mathbf{a} = \mathbf{i} + 3\mathbf{j} - \mathbf{k}$ so $|\mathbf{b} - \mathbf{a}|$

$\sqrt{1^2 + 3^2 + (-1)^2} = \sqrt{11}$

Step 3: Find $|\mathbf{c} - \mathbf{a}|$ where $\mathbf{c} - \mathbf{a} = 3\mathbf{i} - 2\mathbf{j} + 3\mathbf{k}$

so $|\mathbf{c} - \mathbf{a}| = \sqrt{3^2 + (-2)^2 + 3^2} = \sqrt{22}$

Step 4: Find $(\mathbf{b} - \mathbf{a}) \cdot (\mathbf{c} - \mathbf{a})$ using the general result '$\mathbf{a} \cdot \mathbf{b} = a_1b_1 + a_2b_2 + a_3b_3$'

so $= (\mathbf{i} + 3\mathbf{j} - \mathbf{k}) \cdot (3\mathbf{i} - 2\mathbf{j} + 3\mathbf{k}) = 1(3) + 3\,(-2) + (-1)\,(3) =$
$3 - 6 - 3 = -6$

Step 5: Substitute the results from steps 2, 3 and 4 into the formula in step 1 so

$$\cos\theta = \frac{(\mathbf{b} - \mathbf{a}) \cdot (\mathbf{c} - \mathbf{a})}{|\mathbf{b} - \mathbf{a}||\mathbf{c} - \mathbf{a}|} = \frac{-6}{\sqrt{11}\sqrt{22}} = -0.3857$$

$\Rightarrow \theta = 112.69°$

Therefore the angle between $\overrightarrow{AB}$ and $\overrightarrow{AC}$ is 112.69°

c) The area of the triangle $ABC = \frac{1}{2}AB \times AC \times \sin\theta$

Using the results from parts a) and b) $|\overrightarrow{AB}| = \sqrt{11}$ and $|\overrightarrow{AC}| = \sqrt{22}$

so the area of the triangle $ABC = \frac{1}{2} \times \sqrt{11} \times \sqrt{22} \times \sin 112.69°$

$= 7.18$ square units (to 2 d.p.)

Essential notes

AB is the length of $\overrightarrow{AB}$ so $AB = |\overrightarrow{AB}|$. This was explained earlier in the chapter.

Example

Two straight lines have the vector equations:

$\mathbf{r} = -3\mathbf{i} + \mathbf{j} + 4\mathbf{k} + s(-\mathbf{i} + 2\mathbf{k})$ and $\mathbf{r} = -2\mathbf{i} + 2\mathbf{j} + 7\mathbf{k} + t(2\mathbf{i} + \mathbf{j} + \mathbf{k})$

where s and t are scalar parameters.

a) Find the coordinates of the point of intersection of the two lines.

b) Find the angle between the two lines.

Answer

a) Let P, with position vector $\mathbf{r}$, be the point of intersection of the two lines. Where they intersect we can equate the position vectors on each line so

$-3\mathbf{i} + \mathbf{j} + 4\mathbf{k} + s(-\mathbf{i} + 2\mathbf{k}) = -2\mathbf{i} + 2\mathbf{j} + 7\mathbf{k} + t(2\mathbf{i} + \mathbf{j} + \mathbf{k})$

Equating coefficients of $\mathbf{i}$, $\mathbf{j}$ and $\mathbf{k}$ leads to three simultaneous equation:

$$-3 - s = -2 + 2t \qquad (1)$$

$$1 = 2 + t \qquad (2)$$

$$4 + 2s = 7 + t \qquad (3)$$

From equation (2), $t = -1$

Substituting this value for t into equation (1) gives $s = 1$

We must check that equation (3) holds with these value of s and t substituted so work out $4 + 2s$ which is 6 and $7 + t$ which is also 6 so equation 3 is true.

We can now find the point of intersection by substituting for $s = 1$ in $\mathbf{r} = -3\mathbf{i} + \mathbf{j} + 4\mathbf{k} + s(-\mathbf{i} + 2\mathbf{k})) \Rightarrow \mathbf{r} = -4\mathbf{i} + \mathbf{j} + 6\mathbf{k}$ or we obtain the same answer by substituting for $t = -1$ in $\mathbf{r} = -2\mathbf{i} + 2\mathbf{j} + 7\mathbf{k} + t(2\mathbf{i} + \mathbf{j} + \mathbf{k})$
$\Rightarrow \mathbf{r} = -4\mathbf{i} + \mathbf{j} + 6\mathbf{k}$

Therefore the position vector of the point of intersection P is $\mathbf{r} = -4\mathbf{i} + \mathbf{j} + 6\mathbf{k}$ so the coordinates of the point of intersection of the two lines are $(-4, 1, 6)$

b) The angle between the two lines is θ where $\cos\theta = \dfrac{\mathbf{d}\cdot\mathbf{e}}{|\mathbf{d}||\mathbf{e}|}$ and $\mathbf{d}$ and $\mathbf{e}$ are direction vectors of the lines.

For line 1 $\mathbf{r} = -3\mathbf{i} + \mathbf{j} + 4\mathbf{k} + s(-\mathbf{i} + 2\mathbf{k})$ so $\mathbf{d} = -\mathbf{i} + 2\mathbf{k}$

For line 2 $\mathbf{r} = -2\mathbf{i} + 2\mathbf{j} + 7\mathbf{k} + t(2\mathbf{i} + \mathbf{j} + \mathbf{k})$
so $\mathbf{e} = 2\mathbf{i} + \mathbf{j} + \mathbf{k}$

$\mathbf{d}\cdot\mathbf{e} = (-\mathbf{i} + 2\mathbf{k})\cdot(2\mathbf{i} + \mathbf{j} + \mathbf{k}) = -2 + 0 + 2 = 0$ and hence $\cos\theta = 0$

Therefore the two lines are perpendicular and the angle between them is 90°

Method notes

Equal vectors in two dimensions were discussed earlier in the chapter. We can now extend this to three dimensions so if two vectors are equal then their respective **i**, **j** and **k** components are equal.

Essential notes

The scalar product of perpendicular lines = 0 and this was covered earlier in the chapter.

Skew lines

In the last example the two lines had a point of intersection. For many lines this is not the case.

When answering geometric questions about two lines in two dimensions, we have two possibilities, either the two lines have a point of intersection or the two lines are parallel. The method of solution is to solve two simultaneous equations and if there is a solution to both equations then there is an intersection point. If there is no solution to both equations the lines are parallel.

In three dimensions the method of solution is to solve three simultaneous equations with two unknowns, but we might not be able to solve the equations uniquely, in other words we may not find a solution which fits all three equations.

When this happens it means that

- the two lines do not intersect, or
- the two lines are in parallel planes.

In three dimensions, straight lines that do not intersect are called **skew lines**.

The following example illustrates how to decide whether lines are skew.

Essential notes

Imagine two aircraft flying at different heights. The vapour trails seen on the ground are skew lines. They look like they might intersect but there is no danger of collision!

Example

Show that the two straight lines with the vector equations

$\mathbf{r} = -3\mathbf{i} + \mathbf{j} + 4\mathbf{k} + s(-\mathbf{i} + 3\mathbf{j} + 2\mathbf{k})$ and $\mathbf{r} = -2\mathbf{i} + 2\mathbf{j} + \mathbf{k} + t(2\mathbf{i} - 3\mathbf{j} + \mathbf{k})$

where s and t are scalar parameters, are skew lines.

Answer

Step 1: Assume the lines do intersect and equate the position vectors of the point of intersection on each line so

$$-3\mathbf{i} + \mathbf{j} + 4\mathbf{k} + s(-\mathbf{i} + 3\mathbf{j} + 2\mathbf{k}) = -2\mathbf{i} + 2\mathbf{j} + \mathbf{k} + t(2\mathbf{i} - 3\mathbf{j} + \mathbf{k})$$

Step 2: Equate the coefficients of **i**, **j** and **k** from step 1 so we have the three simultaneous equation:

$$-3 - s = -2 + 2t \quad (1)$$

$$1 + 3s = 2 - 3t \quad (2)$$

$$4 + 2s = 1 + t \quad (3)$$

Continued on the next page

Step 3: Solve equations (1) and (2) for s and t.

$(2) + 3 \times (1) \Rightarrow -8 = -4 + 3t \Rightarrow t = -\frac{4}{3}$

and substitute this value for t into (2) $\Rightarrow 1 + 3s = 2 + 4 = 6 \Rightarrow s = \frac{5}{3}$

Step 3: Substitute for s and t into equation (3):

LHS: $4 + 2s = 4 + 2 \times \frac{5}{3} = \frac{22}{3}$

RHS: $1 + t = 1 + \left(-\frac{4}{3}\right) = -\frac{1}{3}$

Which shows that equation (3) cannot be satisfied by these values of s and t so there is no point of intersection of the two lines.

Therefore the lines are skew lines.

Stop and think answers

1. *Given* $\mathbf{a} = 6\mathbf{i} + 8\mathbf{j}$ *then* $|\mathbf{a}| = \sqrt{6^2 + 8^2} = 10$. *We require a unit vector in the same direction as a so it must be of length* $1 \Rightarrow$ *unit vector is* $\frac{1}{10}(6\mathbf{i} + 8\mathbf{j}) = 0.6\mathbf{i} + 0.8\mathbf{j}$.

2. a) *Given* $\mathbf{a} = 3\mathbf{i} + 4\mathbf{j}$ *then* $|\mathbf{a}| = \sqrt{3^2 + 4^2} = 5$
   and if $\mathbf{b} = 9\mathbf{i} + 12\mathbf{j}$ *then* $|\mathbf{b}| = \sqrt{9^2 + 12^2} = 15$

   To find the angle between two vectors use the scalar product so

   $\mathbf{a} \cdot \mathbf{b} = |\mathbf{a}||\mathbf{b}| \cos\theta$ *and* $\mathbf{a} \cdot \mathbf{b} = 3 \times 9 + 4 \times 12 = 75$
   therefore
   $75 = 5 \times 15 \cos\theta \Rightarrow 1 = \cos\theta$ *therefore the angle between them is* $0°$

   This means that the two vectors are parallel.

   We could also have shown this by rewriting $\mathbf{b} = 3(3\mathbf{i} + 4\mathbf{j})$
   So $\mathbf{b} = 3\mathbf{a}$ *which is a scaling of a therefore the two vectors are in the same direction but of different lengths so they are parallel vectors.*

   b) $\mathbf{i} \cdot \mathbf{i} = |\mathbf{i}|\,|\mathbf{i}| \cos 0 = 1$ *as* $\mathbf{i}$ *is a unit vector therefore* $|\mathbf{i}| = 1$ *and* $\cos 0 = 1$

   Similarly $\mathbf{j} \cdot \mathbf{j} = 1$

3. *To find the angle between two lines we find the angle between their direction vectors.*

   Given the line $\mathbf{r} = \mathbf{i} + 5\mathbf{j} + t(-2\mathbf{i} + \mathbf{j})$ *then let the direction vector be* $\mathbf{d} = -2\mathbf{i} + \mathbf{j}$ *and* $|\mathbf{d}| = \sqrt{5}$

   Given the line $\mathbf{r} = 3\mathbf{i} - 4\mathbf{j} + s(-2\mathbf{i} + 5\mathbf{j})$ *then let the direction vector be* $\mathbf{e} = -2\mathbf{i} + 5\mathbf{j}$ *and* $|\mathbf{e}| = \sqrt{29}$

   Using the scalar product $\mathbf{d} \cdot \mathbf{e} = |\mathbf{d}||\mathbf{e}| \cos\theta$ *gives*

   $(-2)((-2) + (1)(5) = \sqrt{5}\,\sqrt{29} \cos\theta$

   so $\frac{9}{\sqrt{5\sqrt{29}}} = \cos\theta \Rightarrow \theta = 41.63°$

   therefore the obtuse angle between the lines is $(180 - 41.63)° = 138.37°$

Questions

You may use a calculator.

1. a) Express $f(x) = \dfrac{3(x+1)}{(x+2)(x-1)}$ in partial fraction form. (3)

   b) Hence, or otherwise, find a series expansion for $f(x)$ in ascending powers, up to and including the term in x^2. For what interval of x is your series expansion valid? (6)

   c) Use your partial fractions in a) to find the exact value of

   $\displaystyle\int_2^5 \frac{3(x+1)}{(x+2)(x-1)}\,dx$, giving your answer as a single logarithm. (5)

2. A curve has parametric equations

   $$x = 3\sin t,\ y = \cos 2t, \qquad 0 \leq t \leq \pi$$

   a) Find an expression for $\dfrac{dy}{dx}$ in terms of the parameter t. (4)

   b) Find the equation of the tangent at the point where $t = \dfrac{\pi}{3}$ (4)

   c) Find a Cartesian equation of the curve in the form $y = f(x)$. (4)

3. A curve C has equation

   $$5x^2 - 4y^2 + 3x - 2y + 6 = 0$$

   Find the equation of the normal to the curve C at the point (0, 1) giving your answer in the form $ax + by + c = 0$ where a, b and c are integers. (7)

4. Use the substitution $u = 3^x$ to find the exact value of $\displaystyle\int_0^1 \frac{3^x}{(3^x+2)^2}\,dx$ (6)

5. The trapezium rule is to be used to approximate the integral

   $$\int_0^2 e^{-x^2}\,dx$$

   a) Complete the table with the missing values of y.

x	0	0.5	1.0	1.5	2.0
y	1		e^{-1}		e^{-4}

   (2)

   b) Use the trapezium rule with five y values to approximate the integral

   $$\int_0^2 e^{-x^2}\,dx$$ (4)

6. The figure shows the graph of the curve with equation

   $y = x\,e^x$ where $0 \le x \le 1$

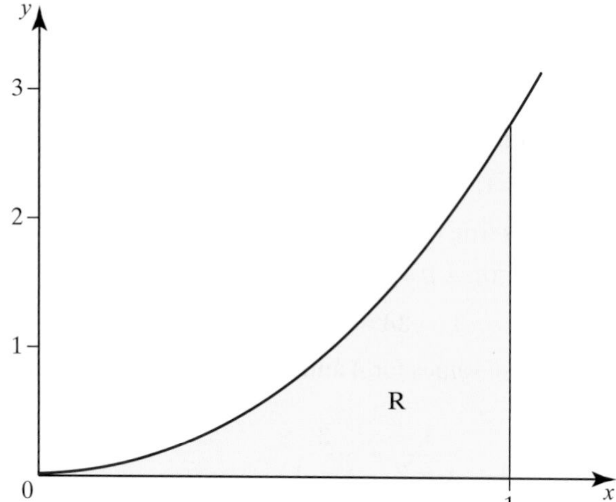

   a) Find the exact value of the area of region R bounded by the line $x = 1$ the x-axis and the curve. (5)

   The region R is rotated through 2π radians about the x-axis.

   b) Find the volume of the solid generated. (6)

7. a) Find, in the form $V = f(t)$, the particular solution of the differential equation:

   $$\frac{dV}{dt} = 4 - 3V$$

   given that $V = 2$ when $t = 0$ (7)

   b) Find the limiting value of V as $t \to \infty$ (1)

8. Two points A and B have position vectors $(6\mathbf{i} + 19\mathbf{j} - \mathbf{k})$ and $(5\mathbf{i} + 15\mathbf{j} + \mathbf{k})$ respectively.

   a) Find, in vector form, an equation of the line l_1 which passes through A and B. (2)

   The line l_2 has equation $\mathbf{r} = 4\mathbf{i} + 9\mathbf{j} - 7\mathbf{k} + t(\mathbf{i} + 5\mathbf{j} + 3\mathbf{k})$ where t is a scalar parameter.

   b) Show that point A lies on the line l_2. (2)

   c) Find the acute angle between the lines l_1 and l_2. (4)

   The point C with position vector $(-11\mathbf{j} - 19\mathbf{k})$ lies on the line l_2.

   d) Find the area of triangle BAC. (4)

Answers

1. a) Let $\dfrac{3(x+1)}{(x+2)(x-1)} \equiv \dfrac{A}{x+2} + \dfrac{B}{x-1}$ (1)

and rewrite with the common denominator $(x+2)(x-1)$

so $\dfrac{3(x+1)}{(x+2)(x-1)} \equiv \dfrac{A(x-1)+B(x+2)}{(x+2)(x-1)}$

Equate the numerators of the fractions so

$3(x+1) \equiv A(x-1) + B(x+2)$

This identity is true for all values of x so

let $x = 1 \Rightarrow 6 = 3B \Rightarrow B = 2$ (1)

and let $x = -2 \Rightarrow -3 = -3A \Rightarrow A = 1$ (1)

Substitute these values for A and B into the original identity which gives

$$\frac{3(x+1)}{(x+2)(x-1)} = \frac{1}{x+2} + \frac{2}{x-1}$$

b) Hence means we can use the result from part a) so

$$\frac{1}{x+2} = (2+x)^{-1} = \tfrac{1}{2}\left(1+\frac{x}{2}\right)^{-1}$$

$$= \frac{1}{2}\left(1 - \frac{x}{2} + \left(\frac{x}{2}\right)^2 + \ldots\right) = \frac{1}{2} - \frac{1}{4}x + \frac{1}{8}x^2 \quad (2)$$

which is valid for $\left|\dfrac{x}{2}\right| < 1$ and similarly

$$\frac{2}{x-1} = -2(1-x)^{-1} = -2(1+x+x^2+\ldots) \quad (2)$$

which is valid for $|x| < 1$ therefore combining the two and including powers of x up to x^2 gives

$$\frac{1}{x+2} + \frac{2}{x-1} \approx \frac{1}{2} - \frac{1}{4}x + \frac{1}{8}x^2 - 2 - 2x - 2x^2$$

$$\approx -\frac{3}{2} - \frac{9}{4}x - \frac{15}{8}x^2 \quad (1)$$

The individual series expansions are valid for $\left|\dfrac{x}{2}\right| < 1$ and $|x| < 1$ so the final result is valid for $\{x: -1 < x < 1\}$. (1)

c) From part a) $\displaystyle\int_2^5 \frac{3(x+1)}{(x+2)(x-1)}\,dx = \int_2^5 \left(\frac{1}{x+2} + \frac{2}{x-1}\right)dx$ (2)

and using integrals of standard functions

$= \left[\ln(x+2)\right]_2^5 + 2\left[\ln(x-1)\right]_2^5 = (\ln 7 - \ln 4) + 2(\ln 4 - \ln 1)$ (2)

which using rules of logarithms simplifies to give

$\ln 7 + \ln 4 = \ln 28$ (1)

2. a) Given the parametric equations

$x = 3\sin t,\ y = \cos 2t$ where $0 \le t \le \pi$ then $\dfrac{dx}{dt} = 3\cos t$

and $\dfrac{dy}{dt} = -2\sin 2t$ so to find $\dfrac{dy}{dx}$ use (2)

$$\frac{dy}{dx} = \frac{\frac{dy}{dt}}{\frac{dx}{dt}} = \frac{-2\sin 2t}{3\cos t} = \frac{-4\sin t\cos t}{3\cos t} = -\frac{4}{3}\sin t \quad (2)$$

because $\sin 2t = 2\sin t\cos t$ from the double angle formula.

b) If $t = \dfrac{\pi}{3} \Rightarrow \dfrac{dy}{dx} = -\dfrac{4}{3}\sin\dfrac{\pi}{3} = -\dfrac{2\sqrt{3}}{3}$ as $\sin\dfrac{\pi}{3} = \dfrac{\sqrt{3}}{2}$ (1)

also if $t = \dfrac{\pi}{3} \Rightarrow x = 3\sin\dfrac{\pi}{3} = \dfrac{3\sqrt{3}}{2}$

and $y = \cos\dfrac{2\pi}{3} = -\dfrac{1}{2}$ because the cosine of an angle is (1)

negative in the second quadrant and $\dfrac{2\pi}{3} = 120^\circ$. $\left(\dfrac{3\sqrt{3}}{2}, -\dfrac{1}{2}\right)$ is a point on the tangent therefore the equation of the tangent is given

by $\dfrac{y - y_1}{x - x_1} = m$ where m is the gradient of the tangent to the curve

and (x_1, y_1) is a point on the tangent so $\dfrac{y + \frac{1}{2}}{x - \frac{3\sqrt{3}}{2}} = -\dfrac{2\sqrt{3}}{3}$ (1)

Simplifying this gives $y + \dfrac{1}{2} = -\dfrac{2\sqrt{3}}{3}\left(x - \dfrac{3\sqrt{3}}{2}\right)$ therefore

$y + \frac{1}{2} = -\dfrac{2\sqrt{3}}{3}x + 3$

and so $y = -\dfrac{2\sqrt{3}}{3}x + \dfrac{5}{2}$ is the equation of the tangent at the point

where $t = \dfrac{\pi}{3}$ (1)

c) Given $x = 3\sin t,\ y = \cos 2t$ to find the Cartesian equation of the curve means we need to eliminate the parameter t between the two equations. (1)

Using $\cos 2t = 1 - 2\sin^2 t$ then $y = \cos 2t = 1 - 2\sin^2 t$ (2)

$= 1 - 2\left(\dfrac{x}{3}\right)^2 \Rightarrow y = 1 - \dfrac{2x^2}{9}$ which is the Cartesian equation of the curve. (1)

3. $5x^2 - 4y^2 + 3x - 2y + 6 = 0$

To find the equation of the normal to the curve at the point (0, 1) we need to find the gradient of the normal (m) at this point then use the equation of a straight line: $\frac{y - y_1}{x - x_1} = m$

$\frac{dy}{dx}$ is the gradient of the tangent at any point so $\frac{-1}{\frac{dy}{dx}}$ gives the gradient of the normal as the tangent and normal are perpendicular to each other.

To find $\frac{dy}{dx}$ we need to differentiate each term of $5x^2 - 4y^2 + 3x - 2y + 6 = 0$ implicitly with respect to x, (remembering that $4y^2$ is a function of a function of x so we use the chain rule to differentiate $4y^2$)

$\Rightarrow 10x - 8y\frac{dy}{dx} + 3 - 2\frac{dy}{dx} = 0$ (3)

Rearranging algebraically gives $\frac{dy}{dx} = \frac{10x + 3}{8y + 2}$ (1)

At the point (0, 1) $\frac{dy}{dx} = \frac{10x + 3}{8y + 2} = \frac{3}{10}$

So the slope of the tangent at (0, 1) is $\frac{3}{10}$

$\Rightarrow$ the slope of the normal at (0, 1) is $-\frac{10}{3}$ (1)

Therefore using $\frac{y - y_1}{x - x_1} = m$ with $x_1 = 0$ and $y_1 = 1$

the equation of the normal is $\frac{y - 1}{x} = -\frac{10}{3}$

$\Rightarrow 3y - 3 = -10x$ (2)

$= 10x + 3y - 3 = 0$

4. To evaluate $\int_0^1 \frac{3^x}{(3^x + 2)^2} dx$ use the method of integration by substitution. (1)

Given $u = 3^x \Rightarrow \frac{du}{dx} = 3^x \times \ln 3$

if $x = 0$, $u = 1$ and if $x = 1$ $u = 3$

Rewrite the integral using the chain rule for

$\int_0^1 dx = \int_1^3 \frac{dx}{du} du$ and substituting for x in terms of u

So $\int_0^1 \frac{3^x}{(3^x + 2)^2} dx = \int_1^3 \frac{1}{(u + 2)^2} \times \frac{du}{\ln 3} = \frac{1}{\ln 3}\int_1^3 \frac{1}{(u + 2)^2} du$ (2)

and rewriting $\frac{1}{(u+2)^2}$ as $(u+2)^{-2}$ before using the rule of integration gives

$$\frac{1}{\ln 3}\int_1^3 \frac{1}{(u+2)^2}du = \frac{1}{\ln 3}\left[\frac{-1}{(u+2)}\right]_1^3$$

which by using the upper and lower limits

evaluates to give $\frac{1}{\ln 3}\left[\frac{-1}{5} - \frac{-1}{3}\right] = \frac{2}{15\ln 3}$ which is the exact value. (3)

5. a) Given $y = e^{-x^2}$:

- when $x = 0.5 = \frac{1}{2}\ y = e^{-\frac{1}{4}}$
- when $x = 1.5 = \frac{3}{2}\ y = e^{-\frac{9}{4}}$

so the completed table is

x	0	0.5	1.0	1.5	2.0
y	1	$e^{-\frac{1}{4}}$	e^{-1}	$e^{-\frac{9}{4}}$	e^{-4}

(2)

b) Using the trapezium rule with 4 strips (5 y values)

$$\int_0^2 e^{-x^2}dx \approx \frac{h}{2}[y_0 + 2(y_1 + y_2 + y_3) + y_4]$$

so $h = \frac{2-0}{4} = 0.5$ therefore (1)

$$\int_0^2 e^{-x^2}dx = \frac{0.5}{2}[1 + 2(e^{-\frac{1}{4}} + e^{-1} + e^{-\frac{9}{4}}) + e^{-4}]$$

$= 0.8806$ (to 4 d.p.) (3)

6. a) Area of R $= \int_0^1 y\,dx$ but $y = xe^x$ and this is a product of two functions of x so we use the method of integration by parts. (1)

Let $u = x \Rightarrow \frac{du}{dx} = 1$ and let $\frac{dv}{dx} = e^x$ so $v = e^x$ therefore

$$\int_0^1 xe^x dx = \left[xe^x\right]_0^1 - \int_0^1 e^x dx = \left[xe^x\right]_0^1 - \left[e^x\right]_0^1 \quad (2)$$

$= [e^1 - 0] - [e^1 - e^0]$

$= 1$ which is the exact value of the area of R. (2)

b) The region R is rotated through 2π radians about the x axis so

volume of revolution $= \int_0^1 \pi y^2 dx$

where $y = xe^x$ so $y^2 = x^2\,(e^x)^2 = x^2\,e^{2x}$ (2)

$\int_0^1 \pi y^2 dx = \int_0^1 \pi x^2 e^{2x} dx$ and this is a product of two functions of x so we use the parts formula with

$u = x^2 \Rightarrow \frac{du}{dx} = 2x$ and $\frac{dv}{dx} = e^{2x}$ so $v = \frac{e^{2x}}{2}$ so

$$\int_0^1 \pi y^2 dx = \int_0^1 \pi x^2 e^{2x} dx = \pi\left[\frac{x^2 e^{2x}}{2}\right]_0^1 - \pi \int_0^1 xe^{2x} dx \qquad (2)$$

and to evaluate $\int_0^1 xe^{2x} dx$ we use the parts formula again with $u = x$

$\Rightarrow \quad \frac{du}{dx} = 1$ and $\frac{dv}{dx} = e^{2x} \Rightarrow v = \frac{e^{2x}}{2}$

$$\text{so } \pi \int_0^1 xe^{2x} dx = \pi\left[\frac{xe^{2x}}{2}\right]_0^1 - \pi \int_0^1 \frac{e^{2x}}{2} dx$$

$$= \pi\left[\frac{xe^{2x}}{2}\right]_0^1 - \pi\left[\frac{e^{2x}}{4}\right]_0^1$$

so volume of solid generated $= \pi\left[\frac{x^2 e^{2x}}{2} - \frac{xe^{2x}}{2} + \frac{e^{2x}}{4}\right]_0^1$

$$\pi\left[\frac{e^2}{4} - \frac{e^0}{4}\right] = \frac{\pi e^2}{4} - \frac{\pi}{4} \qquad (2)$$

7. a) Given $\frac{dV}{dt} = 4 - 3V$ this is a first order differential equation in terms of the variables V and t: to solve we need to find V where $V = f(t)$. (1)

Use the method of separating the variables and integrating and as $4 - 3V = -1(3V - 4)$ then

$$\int \frac{1}{(3V - 4)} dV = \int -dt \qquad (1)$$

and using the standard integral result for ln $(3V - 4) \Rightarrow$

$$\frac{1}{3}\ln(3V - 4) = -t + c \qquad (1)$$

which is a general solution where c is a constant of integration.

Given the boundary conditions $V = 2$ when $t = 0$ if we substitute in these values then

$$\frac{1}{3}\ln(6 - 4) = -0 + c \Rightarrow c = \frac{1}{3}\ln 2 \qquad (1)$$

Substituting for c into the general solution gives

$$\frac{1}{3}\ln(3V - 4) = -t + \frac{1}{3}\ln 2$$

Rearranging algebraically this gives

$$\ln(3V-4) - \ln 2 = -3t \Rightarrow \ln\left(\frac{3V-4}{2}\right) = -3t$$

Writing the equivalent statement gives $3V = 4 + 2e^{-3t}$

$\Rightarrow V = \frac{4}{3} + \frac{2}{3}e^{-3t}$ which is the particular solution. (3)

b) From part a) as $t \rightarrow \infty$, $\frac{2}{3}e^{-3t} \rightarrow 0$ so $V \rightarrow \frac{4}{3}$

which is the limiting value of V. (1)

8. a) If the line l_1 passes through the point A with position vector $6\mathbf{i} + 19\,\mathbf{j} - \mathbf{k}$ and the point B with position vector $5\mathbf{i} + 15\mathbf{j} + \mathbf{k}$ then using the vector form for the equation of the line $\mathbf{r} = \mathbf{a} + \lambda(\mathbf{b} - \mathbf{a})$ where λ is a scalar parameter, $\mathbf{a}$ is the position vector of the point A and $\mathbf{b}$ is the position vector of the point B then the equation of l_1 is $\mathbf{r} = 6\mathbf{i} + 19\mathbf{j} - \mathbf{k} + \lambda((5\mathbf{i} + 15\mathbf{j} + \mathbf{k}) - (6\mathbf{i} + 19\mathrm{j} - \mathbf{k}))$

$\mathbf{r} = 6\mathbf{i} + 19\mathbf{j} - \mathbf{k} + \lambda(-\mathbf{i} - 4\mathbf{j} + 2\mathbf{k})$ (2)

b) Given that $\mathbf{r} = 4\mathbf{i} + 9\mathbf{j} - 7\mathbf{k} + t(\mathbf{i} + 5\mathbf{j} + 3\mathbf{k})$ is the equation of the line l_2 then rearranging this gives

$\mathbf{r} = (4 + t)\mathbf{i} + (9 + 5t)\mathbf{j} + (-7 + 3t)\mathbf{k}$ which means that if the point A with position vector $\mathbf{a} = 6\mathbf{i} + 19\mathbf{j} - \mathbf{k}$ is on this line then the **i, j** and **k** components must be equal so

$4 + t = 6$

$9 + 5t = 19$

$-7 + 3t = -1$

If $4 + t = 6$ then $t = 2$, so $9 + 5(2) = 19$ and $-7 + 3(2) = -1$ which are all true so the point A does lie on line l_2. (2)

c) From part a) the equation of line l_1 is

$\mathbf{r} = 6\mathbf{i} + 19\mathbf{j} - \mathbf{k} + \lambda(-\mathbf{i} - 4\mathbf{j} + 2\mathbf{k})$. If we compare this with the general vector equation of a line $\mathbf{r} = \mathbf{a} + \lambda\mathbf{d}$ then the direction vector of this line is $\mathbf{d} = -\mathbf{i} - 4\mathbf{j} + 2\mathbf{k}$ or we can say the line is parallel to

$\mathbf{d} = -\mathbf{i} - 4\mathbf{j} + 2\mathbf{k}$ (1)

We are given that the equation of line l_2 is $\mathbf{r} = 4\mathbf{i} + 9\mathbf{j} - 7\mathbf{k}) + t(\mathbf{i} + 5\mathbf{j} + 3\mathbf{k})$. If we compare this with the general vector equation of a line we can say that $\mathbf{e} = \mathbf{i} + 5\mathbf{j} + 3\mathbf{k}$ is the direction vector of this line or we can say that the line is parallel to $\mathbf{i} + 5\mathbf{j} + 3\mathbf{k}$ (1)

Therefore the angle between the two lines will be the angle between their direction vectors so we use the scalar product $\mathbf{d} \cdot \mathbf{e} = |\mathbf{d}||\mathbf{e}| \cos\theta$ where θ is the angle between the two vectors $\mathbf{d}$ and $\mathbf{e}$ (1)

$$\cos\theta = \frac{(-\mathbf{i} - 4\mathbf{j} + 2\mathbf{k}) \cdot (\mathbf{i} + 5\mathbf{j} + 3\mathbf{k})}{|-\mathbf{i} - 4\mathbf{j} + 2\mathbf{k}| \times |\mathbf{i} + 5\mathbf{j} + 3\mathbf{k}|}$$

$$= \frac{-1 - 20 + 6}{\sqrt{21} \times \sqrt{35}} = \frac{-15}{\sqrt{21} \times \sqrt{35}} = -0.5533$$

$\theta = 123.6°$(from calculator)

so the acute angle between the lines l_1 and l_2 (1)

$= 180° - 123.6° = 56.4°$

d) Using the general formula for the area of triangle (1)

area of triangle $BAC = \frac{1}{2} AB \times AC \sin\theta$

since A and C both lie on the line l_2, B is a point on l_1 and θ is the angle between AB and AC. (1)

Using the triangle law of addition with position vectors from the origin O, $\overrightarrow{AB} = \overrightarrow{AO} + \overrightarrow{OB}$ so $\overrightarrow{AB} = -\mathbf{a} + \mathbf{b}$

So $AB = |\mathbf{b} - \mathbf{a}| = |-\mathbf{i} - 4\mathbf{j} + 2\mathbf{k}| = \sqrt{21}$

Similarly $AC = |\mathbf{c} - \mathbf{a}| = |-6\mathbf{i} - 30\mathbf{j} - 18\,\mathbf{k}| = \sqrt{1260}$

Therefore area of triangle $BAC = \frac{1}{2}\sqrt{21}\sqrt{1260} \sin 56.4°$ (2)

$= 67.7$ square units

Set notation

$\in$	is an element of
$\notin$	is not an element of
$\{x_1, x_2, \ldots\}$	the set with elements $x_1, x_2, \ldots$
$\{x: \ldots\}$	the set of all x such that ...
$\mathrm{n}(A)$	the number of elements in set A
$\varnothing$	the empty set
ε	the universal set
A'	the complement of the set A
$\mathbb{N}$	the set of natural numbers, $\{1, 2, 3, \ldots\}$
$\mathbb{Z}$	the set of integers, $\{0, \pm 1, \pm 2, \pm 3, \ldots\}$
$\mathbb{Z}^+$	the set of positive integers, $\{1, 2, 3, \ldots\}$
$\mathbb{Z}_n$	the set of integers modulo n, $\{0, 1, 2, \ldots, n-1\}$
$\mathbb{Q}$	the set of rational numbers, $\left\{\frac{p}{q} : p \in \mathbb{Z}, q \in \mathbb{Z}^+\right\}$
$\mathbb{Q}^+$	the set of positive rational numbers, $\{x \in \mathbb{Q} : x > 0\}$
$\mathbb{Q}_0^+$	the set of positive rational numbers and zero, $\{x \in \mathbb{Q} : x \geq 0\}$
$\mathbb{R}$	the set of real numbers
$\mathbb{R}^+$	the set of positive real numbers $\{x \in \mathbb{R} : x > 0\}$
$\mathbb{R}_0^+$	the set of positive real numbers and zero, $\{x \in \mathbb{R} : x \geq 0\}$
$\mathbb{C}$	the set of complex numbers
(x, y)	the ordered pair x, y
$A \times B$	the cartesian product of sets A and B, ie $A \times B = \{(a, b) : a \in A, b \in B\}$
$\subseteq$	is a subset of
$\subset$	is a proper subset of
$\cup$	union
$\cap$	intersection
$[a, b]$	the closed interval, $\{x \in \mathbb{R} : a \leq x \leq b\}$
$[a, b), [a, b[$	the interval $\{x \in \mathbb{R} : a \leq x < b\}$
$(a, b],]a, b]$	the interval $\{x \in \mathbb{R} : a < x \leq b\}$
$(a, b),]a, b[$	the open interval $\{x \in \mathbb{R} : a < x < b\}$
yRx	y is related to x by the relation R
$y \sim x$	y is equivalent to x, in the context of some equivalence relation

Miscellaneous symbols

$=$	is equal to
$\neq$	is not equal to
$\equiv$	is identical to or is congruent to
$\approx$	is approximately equal to
$\cong$	is isomorphic to
$\propto$	is proportional to
$<$	is less than
$\leq$	is less than or equal to, is not greater than
$>$	is greater than
$\geq$	is greater than or equal to, is not less than
∞	infinity
$p \wedge q$	p and q
$p \vee q$	p or q (or both)
$\sim p$	not p
$p \Rightarrow q$	p implies q (if p then q)
$p \Leftarrow q$	p is implied by q (if q then p)
$p \Leftrightarrow q$	p implies and is implied by q (p is equivalent to q)
$\exists$	there exists
$\forall$	for all

Operations

$a + b$	a plus b
$a - b$	a minus b
$a \times b$, ab, $a.b$	a multiplied by b
$a \div b$, $\frac{a}{b}$, a/b	a divided by b
$\sum_{i=1}^{n} a_i$	$a_1 + a_2 + \ldots + a_n$
$\prod_{i=1}^{n} a_i$	$a_1 \times a_2 \times \ldots \times a_n$
$\sqrt{a}$	the positive square root of a
$\lvert a \rvert$	the modulus of a
$n!$	n factorial
$\binom{n}{r}$	the binomial coefficient $\frac{n!}{r!(n-r)!}$ for $n \in \mathbb{Z}^+$ $\frac{n(n-1)\ldots(n-r+1)}{r!}$ for $n \in \mathbb{Q}$

Functions

$f(x)$	the value of the function f at x
$f: A \to B$	f is a function under which each element of set A has an image in set B
$f: x \to y$	the function f maps the element x to the element y
f^{-1}	the inverse function of the function f
$g \circ f$, gf	the composite function of f and g which is defined by $(g \circ f)(x)$ or $gf(x) = g(f(x))$
$\lim_{x \to a} f(x)$	the limit of $f(x)$ as x tends to a
$\Delta x, \delta x$	an increment of x
$\frac{dy}{dx}$	the derivative of y with respect to x
$\frac{d^n y}{dx^n}$	the nth derivative of y with respect to x
$f'(x), f''(x), \ldots, f(n)(x)$	the first, second, ..., nth derivatives of $f(x)$ with respect to x
$\int y\,dx$	the indefinite integral of y with respect to x
$\int_a^b y\,dx$	the definite integral of y with respect to x between the limits $x = a$ and $x = b$
$\dot{x}, \ddot{x}$	the first, second, ... derivatives of x with respect to t

Exponential and logarithmic functions

e	base of natural logarithms
e^x, $\exp x$	exponential function of x
$\log_a x$	logarithm to the base a of x
lin x, $\log_e x$	natural logarithm of x
$\lg x$, $\log_{10} x$	logarithm of x to base 10

Circular and hyperbolic functions

sin, cos, tan, cosec, sec, cot	the circular functions
arcsin, arccos, arctan, arccosec, arcsec, arccot	the inverse circular functions

Vectors

$\mathbf{a}$	the vector $\mathbf{a}$
$\overrightarrow{AB}$	the vector represented in magnitude and direction by the directed line segment AB
$\hat{a}$	a unit vector in the direction of $\mathbf{a}$
$\mathbf{i}, \mathbf{j}, \mathbf{k}$	unit vectors in the directions of the cartesian coordinate axes
$\lvert\mathbf{a}\rvert, a$	the magnitude of $\mathbf{a}$
$\lvert\overrightarrow{AB}\rvert, AB$	the magnitude of $\overrightarrow{AB}$
$\mathbf{a} \cdot \mathbf{b}$	the scalar product of $\mathbf{a}$ and $\mathbf{b}$.

Formulae you need to remember

Trigonometry

$\cos^2 A + \sin^2 A \equiv 1$

$\sec^2 A \equiv 1 + \tan^2 A$

$\text{cosec}^2 A \equiv 1 + \cot^2 A$

$\sin 2A \equiv 2 \sin A \cos A$

$\cos 2A \equiv \cos^2 A - \sin^2 A$

$$\tan 2A \equiv \frac{2 \tan A}{1 - \tan^2 A}$$

Differentiation

function	**derivative**
$\sin kx$	$k \cos kx$
$\cos kx$	$-k \sin kx$
e^{kx}	ke^{kx}
$\ln x$	$\frac{1}{x}$
$f(x) + g(x)$	$f'(x) + g'(x)$
$f(x)g(x)$	$f'(x)g'(x) + f(x)g'(x)$
$f(g(x))$	$f'(g(x))g'(x)$

Integration

function	**integral**
$\cos kx$	$\frac{1}{k} \sin kx + c$
$\sin kx$	$-\frac{1}{k} \cos kx + c$
e^{kx}	$\frac{1}{k} e^{kr} + c$
$\frac{1}{x}$	$\ln \lvert x \rvert + c, x \neq 0$
$f'(x) + g'(x)$	$f(x) + g(x) + c$
$f'(g(x))g'(x)$	$f(g(x)) + c$

Vectors

$$\begin{pmatrix} x \\ y \\ z \end{pmatrix} \cdot \begin{pmatrix} a \\ b \\ c \end{pmatrix} = xa + yb + zc$$

Formulae given in the formulae booklet

Mensuration

Surface area of sphere $= 4\pi r^2$

Area of curved surface of cone $= \pi r \times$ slant height

Arithmetic series

$u_n = a + (n-1)d$

$S_n = \frac{1}{2}n(a+l) = \frac{1}{2}n[2a + (n-1)d]$

Cosine rule

$a^2 = b^2 + c^2 - 2bc\cos A$

Binomial series

$$(a+b)^n = a^n + \binom{n}{1}a^{n-1}b + \binom{n}{2}a^{n-2}b^2 + \ldots + \binom{n}{r}a^{n-r}b^r + \ldots + b^n \quad (n \in \mathbb{N})$$

where $\binom{n}{r} = {}^n\mathrm{C}_r = \frac{n!}{r!(n-r)!}$

$$(1+x)^n = 1 + nx + \frac{n(n-1)}{1 \times 2}x^2 + \ldots + \frac{n(n-1)\ldots(n-r+1)}{1 \times 2 \times \ldots \times r}x^r + \ldots \quad (|x| < 1, n \in \mathbb{R})$$

Logarithms and exponentials

$\log_a x = \frac{\log_b x}{\log_b a}$

Geometric Series

$u_n = ar^{n-1}$

$S_n = \frac{a(1-r^n)}{1-r}$

$S_\infty = \frac{a}{1-r}$ for $|r| < 1$

Numerical integration

The trapezium rule: $\int_a^b y\,dx \approx \frac{1}{2}h\{(y_0 + y_n) + 2(y_1 + y_2 + \ldots + y_{n-1})\}$, where $h = \frac{b-a}{n}$

Logarithms and exponentials

$e^{x\ln a} = a^x$

Trigonometric identities

$\sin(A \pm B) = \sin A\cos B \pm \cos A\sin B$

$\cos(A \pm B) = \cos A\cos B \mp \sin A\sin B$

$\tan(A \pm B) = \frac{\tan A \pm \tan B}{1 \mp \tan A\tan B} \quad (A \pm B \neq (k + \frac{1}{2})\pi)$

$\sin A + \sin B = 2\sin\frac{A+B}{2}\cos\frac{A-B}{2}$

$$\sin A - \sin B = 2\cos\frac{A+B}{2}\sin\frac{A-B}{2}$$

$$\cos A + \cos B = 2\cos\frac{A+B}{2}\cos\frac{A-B}{2}$$

$$\cos A - \cos B = -2\sin\frac{A+B}{2}\sin\frac{A-B}{2}$$

Differentiation

$\mathbf{f}(x)$	$\mathbf{f}'(x)$
$\tan kx$	$k\sec^2 kx$
$\sec x$	$\sec x\tan x$
$\cot x$	$-\mathrm{cosec}^2\, x$
$\mathrm{cosec}\, x$	$-\mathrm{cosec}\, x\cot x$
$\dfrac{f(x)}{g(x)}$	$\dfrac{f'(x)g(x) - f(x)g'(x)}{(g(x))^2}$

Integration (+ constant)

function	**integral**
$\sec^2 kx$	$\dfrac{1}{k}\tan kx$
$\tan x$	$\ln\lvert\sec x\rvert$
$\cot x$	$\ln\lvert\sin x\rvert$
$\mathrm{cosec}\, x$	$-\ln\lvert\mathrm{cosec}\, x + \cot x\rvert$, $\ln\lvert\tan(\frac{1}{2}x)\rvert$
$\sec x$	$\ln\lvert\sec x + \tan x\rvert$, $\ln\lvert\tan(\frac{1}{2}x + \frac{1}{2}\pi)\rvert$

$$\int u\frac{dv}{dx}\,dx = uv - \int v\frac{du}{dx}\,dx$$